Practical
STONEMASONRY
Made Easy

The Weight of Stone
for Steve Kennedy

The weight of stone is peace
for hands that know
the weightlessness of words:
Four bearded boys, dreamers wise
to what can be told and portable
in weightless lines, we sigh
in the glittering April wind
and curve like dancers
over the hard, true flesh of rock,
lifting with bent, trembling legs
this squarish body heavy as half a car.
It comes alive only when we strain upward,
marvelling at the weakness of our plans.
Struck with a steel hammer,
it gives off the bitter smell of time.
I love the strength of its stillness.
Its patience would be ecstasy in a living thing.
Our touch so light, our hands weak as kisses
on its dreamless forehead,
it holds us here straining upward,
empty-headed and glad of it in the April sun.

—*Will Lane, September 1987*

No. 2915
$24.95

Practical STONEMASONRY Made Easy

Stephen M. Kennedy

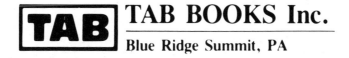

TAB BOOKS Inc.

Blue Ridge Summit, PA

This book is dedicated to the peasants,
who grew the coffee and made this book possible,
and to my family,
for putting up with me

FIRST EDITION

FIRST PRINTING

Copyright © 1988 by TAB BOOKS Inc.

Printed in the United States of America

Library of Congress Cataloging in Publication Data

Kennedy. Stephen M.
    Practical stonemasonry made easy / by Stephen M. Kennedy.
        p.   cm.
    Includes index.
    ISBN 0-8306-1115-0   ISBN 0-8306-2915-7 (pbk.)
    1. Stonemasonry—Amateurs' manuals. I. Title.
TH5411.K46   1988   87-28934
693'.1—dc19   CIP

Questions regarding the content of this book
should be addressed to:

    Reader Inquiry Branch
    TAB BOOKS Inc.
    Blue Ridge Summit, PA 17294-0214

CONTENTS

ACKNOWLEDGMENTS *ix*

INTRODUCTION *xi*

1. THE BEGINNING *1*
   - ☐ The Origin of Stones *1*
   - ☐ Chemical Classifications *2*
   - ☐ What Makes a Stone Desirable for Construction? *3*
   - ☐ Everyday Stones *4*
   - ☐ Philosophy and the Stonemason *5*
   - ☐ Why Build with Stone? *6*
   - ☐ Who Can Build with Stone? *8*
   - ☐ Cost of Building with Stone *9*
   - ☐ When To Build with Stone *9*
   - ☐ Predicting Time *11*
   - ☐ Practical Applications *12*
   - ☐ What Is a Stone Wall? *14*

2. STACKING AND GATHERING STONE *15*
   - ☐ Masonry in Miniature *15*
   - ☐ Build a Miniature Stone Wall *20*
   - ☐ Gathering Stones *22*
   - ☐ Shapes of Stones for Different Jobs *30*

3. LAYOUT AND FOOTINGS *33*
   - ☐ Layout for a Rubble-Stone Footing *33*
   - ☐ Building a Rubble-Stone Footing *39*

4. HOW TO LAY RUBBLE STONE WITHOUT MORTAR *45*
   - ☐ The Basics of Laying Stone Walls *45*
   - ☐ Stacking Imperfectly Shaped Stones *49*
   - ☐ Finer Points of Stone Stacking *55*
   - ☐ Building a Low Retaining Wall *63*

5. SMALL STONE DAMS 73
☐ Think Before You Build a Dam 73
☐ One-Hour Dam 75
☐ Two-Day Dam 77
☐ The Waterfall 81

6. MORTAR 85
☐ Lime and Cement 85
☐ Aggregate 87
☐ Mortar 89
☐ Mixing Mortar 92

7. STONE FOUNDATIONS 97
☐ Stone Piers 97
☐ Building a Complete Stone Foundation 107
☐ Underpinning a Stone Foundation 112

8. STONE FACING 120
☐ Laying Stone Against Block 120
☐ The Dry-Laid Look in a Mortared Facing 127
☐ Facing for Fire Safety 131
☐ Facing a Block Chimney 132

9. CHIMNEYS AND FIREPLACES 137
☐ Wood Burning 137
☐ Break throughs 139
☐ Masonry Fireplaces 140
☐ A Stonemasonry Chimney 141
☐ Fireplaces 149
☐ Parts of a Fireplace 151
☐ Codes for Masonry Fireplaces and Chimneys 155
☐ Troubleshooting for Chimneys and Fireplaces 157

10. STONE STEPS 160
☐ Stone Steps That Follow a Hill 160
☐ Dry-Laid Steps 166
☐ Stone Steps in Mortar 169

11. FLAGSTONE 173
☐ Finding Stone 173
☐ Thin Stone 173
☐ Using Flagstone in Floors 174

□ Flagstone without Mortar *174*
□ Laying Flagstone in Mortar *177*
□ Pointing and Finishing Flagstone *185*
□ Coatings on Floors *186*
□ Laying Flagstone Vertically *186*

12. REPAIRING AND REPOINTING STONEMASONRY *188*

□ Fixing a Blown-Out Retaining Wall *188*
□ Repairing a Collapsed Wall *193*
□ Repointing Stonemasonry *199*

13. SPECIAL ASPECTS OF STONEMASONRY *204*

□ Ashlar Masonry *204*
□ Nonsquare Buildings of Stone *208*
□ Aesthetic Masonry *211*
□ Laying Brick as Rubble Masonry *216*
□ Missing Projects *217*
□ Stone Houses *220*

14. SOLAR HEAT AND STONEMASONRY *223*

□ Approaches to Using Thermal Mass *223*
□ Trombe Walls *225*
□ Some Passive Solar Houses with Stone *226*
□ Kennedy House *239*
□ A Stone Wall of Maximum Practicality *243*
□ In Conclusion *246*

BIBLIOGRAPHY *249*

INDEX *253*

ACKNOWLEDGMENTS

The Drawings: Danne Kennedy
The Photographs: Stephen Kennedy, Danne Kennedy, Will Lane
Project Credits (in order of appearance):

1-4. Todd Mudd
2-9 through 2-11. S. Kennedy
4-34 through 4-37. D. Swope, M. Hieberg, S. Kennedy
4-38 and 4-39. S. Kennedy
7-1. Tom and Diane Sander and friends
7-9. Doug Swope, Candace Kugel, and friends
7-12. Mike Wizer, S. Kennedy
7-20. Candace Kugel and Doug Swope
7-21 through 7-28. Doug Swope, S. Kennedy
8-2 and 8-3. S. Kennedy
8-6. S. Kennedy
8-7 through 8-11. S. Kennedy
8-12 through 8-14. David and Stephen Kennedy
8-15 through 8-17 S. Kennedy

9-7. S. Kennedy
9-8. T. Mudd, S. Kennedy
9-9 and 9-10. S. Kennedy
9-14 and 9-16. T. Mudd, S. Kennedy, Alan Paulson
10-1 and 10-2. Dorothy Lamberti
10-3. S. Kennedy
10-4. S. Kennedy
10-8 and 10-9. D. Kennedy, S. Kennedy
10-10. S. Kennedy, D. Swope
10-11. S. Kennedy
11-5. S. Kennedy
11-6. Danne and S. Kennedy
12-1 through 12-5. D. Swope, S. Kennedy
12-8 through 12-12. Fred Hough, S. Kennedy
13-8 through 13-17. S. Kennedy

13-19. S. Kennedy
13-20. S. Kennedy, Paul Qually
14-1. T. Mudd, A. Paulson, S. Kennedy
14-2 through 14-4. masonry Dave Pickering, S. Kennedy
14-5 and 14-6. T. Mudd, S. Kennedy
14-7. T. Mudd
14-8. contractor Jerry Justice, stonework S. Kennedy
14-9 through 14-11. S. Kennedy
14-12. T. Mudd, S. Kennedy
14-13. stonework T. Mudd, S. Kennedy
14-14 through 14-16. S. Kennedy and friends, notably John Lalima
14-17 and 14-18. designed by S. Kennedy.

INTRODUCTION

THE purpose of this book is to give to the nonprofessional, casual builder the knowledge to build beautiful and lasting structures of uncut stone. The projects that the book unfolds are inexpensive, relatively unskilled, and not terribly time-consuming. Many of the jobs used to illustrate procedures in the book were done by me with a helper who had never done any stonemasonry—usually the owner of the project—and I was able to see the most commonly made errors these helpers would make. For example, if you have a definite space to fill in a stone wall, between two stones, and you walk over to a pile of rocks to find one that looks like it would fit just right, it will be too big.

Beginners tend to make mortar too wet. Perhaps if they know that beforehand, it will come out just right. Novices don't get enough sand to finish the job, and at least one of the bags of cement gets ruined by water. According to my statistics, only 6 percent of those doing a stone project will gather enough stone. The rest think they have enough, but they will need to find more before the job is done. The most frightening statistic is this: no one has ever finished a stonemasonry job as fast as they thought they would.

This book has arrows which point the way toward projects that are likely to be gratifying and practical; and it has caution signs in areas that are likely to be problems. There is much emphasis on beauty here. If beauty doesn't matter, you should lay blocks, not stone. I'm not putting down blocks—they certainly have a useful function—but if your stone wall doesn't come out looking a heck of a lot better than a block wall, you've wasted a lot of effort. Stonemasonry evolved when there were no alternatives. Now that we have concrete and cement blocks, aesthetics has become the main reason to build with stone. This has lead to the look of stone being more important than its structural characteristics. Now there are two types of stonemasonry: the old kind, that incredibly overbuilt conservative time-honored type, and the new kind, a fakery that tells the eye it is looking at a wall of stone, when it isn't.

My bent is to build real structures of stone, but to look for ways to make it easier and more practical. Using *rubble*, or uncut stone, instead of *ashlar*, or cut stone, is the first big step in making stonemasonry affordable. All the projects in this book are of rubble masonry, and many other money- and time-saving details are introduced.

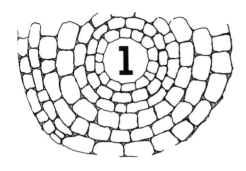

THE BEGINNING

HAVING a general understanding of the diverse circumstances through which stones come into existence might increase your appreciation of the materials you will be working with. Every rock you pick up has been on an incredible journey, and a reverence for that journey will put you in a better frame of mind to cope with the horrible weight that is stonemasonry.

THE ORIGIN OF STONES

A *stone* is a solid piece of nonmetallic mineral substance. The way a stone is formed gives it its geological classification, of which there are three.

Igneous

Igneous stone is formed when magma below the earth's crust comes to the surface and solidifies. It is the great pressure from above that keeps the magma in a liquid condition, but oddly, that liquid pre-stone is less dense than the solid version near the surface, so it tries to float up to the surface of the planet. Volcanoes are localized results of that phenomenon, where the liquid matter spews forth onto the surface of the planet.

Volcanoes are responsible for the release of new minerals and atmospheric gases into the life cycles on earth. Basalt is the most common stone to be formed volcanically, and it is composed mainly of silica (SiO_2), as is trap, gabbro, diabase, and other primordial stone. The other common mineral that is released from below the crust through volcanoes is feldspar, containing alumina and calcium compounds.

Again, stone formed from a liquid condition is called igneous. All other stones are formed from igneous stones.

Metamorphic

Metamorphic stones are created from previously existing materials that change. Feldspar, when exposed to weathering on the surface, becomes clay, which through enormous pressure can become the metamorphic stone called slate. Granite is formed from quartz and feldspar, sometimes at molten temperatures, and can be both igneous and metamorphic. For the most part, metamorphic stones are created by the crushing weight of the parent stone above it. Marble is limestone that has been forced into a tighter crystal arrangement. Quartzite is metamorphic quartz. Gneiss and schist are other common metamorphic stones.

Sedimentary

Sedimentary stones are formed from sediments on the surface of the earth. These rocks cover 75 percent of the crust's surface, but only account for 5 percent of its volume. Shale, sandstone, and limestone make up 99 percent of the sedimentary stones. Where do these sediments originate? The shale comes from clay deposits that were formed from feldspar at the crust. The sandstone comes from river-carried erosion of other stone particles and minerals that build up at the edges of continents. Plates of the crust in slow collision eventually bring this sandstone above sea level. In my area of the country, the sandstone is on the mountaintops, and when it rains hard, tiny pieces of those stones wash down the creek to the ocean, where a new generation of sandstone is being formed. Sandstone also can form from particles of eroded igneous stone.

Limestone is formed mainly from the buildup of exoskeletons of tiny sea animals that drift to the ocean floor when they die. The shell material is mainly calcium carbonate. Some limestone is formed from calcium salts that precipitate out of seawater. Limestone is constantly being quarried for building stone, gravel, cement lime, and agricultural lime. When you lay limestone building stone, you use more limestone in the mortar. You also use sandstone in the mortar. Sand itself is a form of sandstone. A cement block wall is made almost entirely of old stones of one type or another.

It is interesting how we are dependent on stones for our existence. Plants assimilate minerals from weathered stones to assemble their nutrients, and we are directly or indirectly dependent on these nutrients. According to Robert Foster in his textbook *Physical Geology,* an acre of alfalfa, yielding four tons per year, needs two tons of weathered rocks to obtain the necessary phosphorus, calcium, magnesium, potassium, and so on. This is 5 1/2 times more than can be supplied from being released from the stone naturally, hence fertilizer must be used. A pound of steak has two ounces of minerals obtained from the soil by plants. The minerals get into the soil from the weathering of stone. Lichens can live directly on stones and weather their own minerals.

Rocks are very much involved in the life process; in fact, our very bodies are made of bits of former stones. Admittedly, stones are slow, but they aren't out of the picture of continual change. Along with atmospheric gases, water, and subatomic pieces of the sun, they are the building blocks of life on earth.

CHEMICAL CLASSIFICATIONS

Of more concern to the builder than the geological classification of stone is the chemical classification. This is an attempt at organizing stones of every possible sort into three main categories of dominant mineral content.

Siliceous

Siliceous stone is composed mainly of silica, or silicon dioxide (SiO_2). This category includes most of the igneous stone that spews out of volcanoes, like basalt, as well as compressed, cemented sediments of that type of stone, like sandstone. An exceptionally pure silicon dioxide sandstone is white quartz. When it is squashed into a tighter crystal, the meta-

morphic quartzite if formed. The silica of the magma can go through a lot of changes, but all those different forms are still siliceous stones.

Argillaceous

Argillaceous stone is dominated by the presence of alumina (AlO_3). This alumina, with its other compounds of calcium, potassium, etc., originates from feldspar in the crust and magma. When the feldspar minerals come in contact with the atmosphere and water, they change into claylike compounds. Slate is the most common (to the mason) argillaceous stone. It is sort of a petrified clay. When clay gets into other stone in varying quantities, we get into border areas between categories. Brownstone, for example, happens when sand and clay together are washed to the ocean and build up into a softer, earthier sandstone. Brick is an artificial argillaceous stone.

Calcareous

Calcareous stone is dominated by the presence of calcium carbonate ($CaCO_3$), or lime. Lime originates from the bodies of sea creatures, which accumulate at the ocean floor. Limestone is the main calcareous stone. When it is pressurized for millennia, marble is formed. Marble, therefore, is a metamorphic rock, but because it is still mainly lime after its metamorphosis, it is also calcareous in classification.

Between Classifications

Argillaceous limestone has clay impurities. Ironstone is a limestone with clay impurities and iron impurities which exceed the quantity of lime. Chert is limestone with siliceous deposits and flakes. Siliceous limestone has the sand part worked in, as in hornstone. Dolomite is limestone with a third or more magnesium carbonate.

The variations and combinations between classifications are nearly endless, but they are distinguished by the abundance of the glassy (siliceous) stuff, the clayey (argillaceous) stuff, or the limy (calcareous) stuff. The quantities of the different basic compounds, the way in which they were formed, and the presence of other minerals in smaller quantities are what give rocks the particular traits that make them desirable for certain applications.

WHAT MAKES A STONE DESIRABLE FOR CONSTRUCTION?

The strongest common rock is trap, an ancient igneous, siliceous, unstratified primal stone capable of withstanding loads of up to 67,000 pounds per square inch (psi). Gabbro, basalt, and diabase are similar. These stones sometimes show up in masonry, but they are very hard to work, or quarry, because they are not formed in layers and do not lend themselves to cleavages. Granite and related stone are slightly more manageable, and can be cut and shaped. Still, granite is so hard and heavy that normally it is used only when great strength is needed, like on railroad bridge piers. Granite is also great for resisting the effects of the weather. Ashlar granite stonemasonry is serious construction—the bottom line in building material. It is also too expensive for the nonprofessional, do-it-yourselfer.

Ease of Quarrying

Traditionally, what has made stone desirable for masonry is its ease of quarrying. Sedimentary stone is the easiest to remove in large quantities from the ground because it is formed in layers, and can be wedged up in layers with quarry tools. If there was a vein of good limestone at the surface, near an access, it would be quarried, not so much because it was a certain kind of stone, but because it was available.

Certain quarries are noted for a strain of rock that is particularly desirable for building stone because of its workability, its color, its durability, and so on. Brownstone was quarried enthusiastically in Connecticut as early as 1665, according to Charles E. Peterson in his book *Building America* (Chilton Book Co., Radnor Pa.). It was discovered much later that this earthy, easy to shape stone would not stand up to the elements as well as other common stone. Most of us don't think of rocks as things that wear out, but even the hardest granite steps will show deep indentation after a century or two of people walking on them.

In quarries, blasting is not desirable for obtaining building stone because the shocks can set up fissures in the stone. Blasting is used to get junk stone and earth out of the way, but the stone to be used for masonry is split away from the earth with wedges. It is easier to mine wet or green stone, but it must be seasoned (dried out) before use. Stone with quarry sap (water) becomes harder as it cures, much like wood.

I must assume that you are in no position to have special cut sandstone shipped in from 100 miles away. Chances are that you will be working with uncut rubble, or fieldstone. If you're lucky, you might have access to a ruins that includes some rough-dressed stone (FIG. 1-1).

Proximity

These days what makes a particular stone desirable is its proximity, its lack of cost, and then its other characteristics. The exception might be a small stone-facing job where you could afford to be choosy. If you only need about one ton of stones for your project, you might find it worthwhile to hunt for the perfect rocks.

Fig. 1-1. *Rough-dressed stone in a ruins.*

Durability

On an interior stone project, the material needn't be very durable, unless it is supporting a lot of weight. Some stones last such a short while that they aren't worth building with, especially in areas of harsh weather. If you can crumble a stone in your hands or scratch its surface with a fingernail, it isn't fit for masonry.

Absorptiveness

In exterior work, the most important aspect of a particular stone's durability is its absorptiveness. If a stone is at all spongelike, it will soak up water. The water will freeze and rupture the wet stone. Soon, pieces of the stone will flake away.

Even where it doesn't freeze, absorptive stone is damaged by the slightly acidic, corrosive action of the water it takes in. The increasing acidity of the rain is wreaking havoc on some of the old limestone structures in the Northeast United States. Some effort is being made to find materials that can handle the acidity better. I think the limestone masonry was done right and it's the acid rain that must change, although several new companies sell various coatings to protect buildings from the new rain.

You can pretty much tell a stone's absorptiveness by its density. If it has a lot of minute internal spaces, it can take on water. Granite, basalt, and marble soak up almost no water. Limestone absorbs more, sandstone more yet, and bricks and mortar are the most absorptive.

Strength

For houses and small structures, the compressive and shearing strength of stone is hardly ever at issue. The exception might be in lintel stones or shelf stones that extend out of a wall, where good tensile strength would be needed. Usually, stone is no good as a beam. Where it is used as a lintel—over a doorway, for instance—very little weight is actually supported by the lintel stone if the stones above are well knit into the rest of the wall.

Avoid using stone to span over space, except as an arch, or in a case where the space is narrow and there would be very little weight above the bridge stone. If you don't use stone in situations requiring good tensile strength, you needn't worry about the tensile strength of stone.

Great compressive strength is also unnecessary in houses and small buildings, so there would be no reason to choose one type stone over another because the one could support only 10 times more weight than it needed to and the other could support 100 times more weight than required. It would probably be better to have weaker stone that you could shape with a hammer, than have stone that can support awesome loads, but wouldn't yield to the chisel.

Fire Resistance

Fire resistance is another factor that might make one type of stone more durable than another. Granite, which is so strong, falls apart badly in a fire. It is composed of minerals that have different rates of expansion, and as the temperature rises, the rock breaks apart. Sandstone fares much better in a fire. Soapstone (which is a compressed form of talc) is the only stone that can handle temperature changes well enough to be considered as a material for stoves.

Basically, the fire resistance of some stone over another kind would not be enough reason to get that kind. If the house burns down, but the stonework remains intact, it still might not be a good idea to rebuild on the old stone. Most any type of stone house can handle the heat of a small fire that is extinguished. If you avoid using stone where great fire resistance is needed—as in the combustion chamber of a fireplace—then the fire resistance of different types of stone is irrelevant. There's no sense making a house of soapstone so it can survive being burned down a lot.

EVERYDAY STONES

Although there are many varieties of stone, each possessing different features, what is much more important to the mason is the shape of the stone (are they remotely squarish?), the proximity (are they within a few miles?), and the price (can I get them for free?). The other factors are considerations from a bygone era.

If you are after a practical experience with stonemasonry, chances are you will be working with rubble sandstone or limestone. If I had to choose, I'd choose both, although traditional stonemasonry frowns on mixing stone types in a job. The shape of the stones and the way they are stacked will have a lot more to do with the longevity of a building than will the composition of the stones themselves.

Building codes in Chicago near the turn of the century, set by eminent architects and engineers, included the following interesting safety codes for different types of stonemasonry: Uncoursed limestone rubble laid in portland cement mortar was good for 100 psi load. The same stone, laid in courses could accept a load of 200 psi. Ashlar, or cut limestone, laid in courses, in portland cement mortar, was good for 400 psi. The point of this information is that a wall of stone achieves a big jump in strength simply by the way it is stacked, and another big jump in strength if the individual units are squared to fit.

You can't have the stone cut very easily, but you can lay stone in courses with no extra cost. If you can find fieldstone that is shaped right, you will be able to build a much stronger wall than if you are working with unselected rubble. The emphasis in this book is on obtaining the best wall through a selection of well-shaped stones and by a certain method of assemblage.

Calculations have been made on the cost of rubble vs. ashlar on large stone buildings near the year 1900. A cubic foot of rubble-stone wall cost about $.20 assembled, while a cubic foot of fine finished granite ashlar cost $3.00. It took the stone cutter between 5 and 6 days to prepare the stones for a mere cubic yard of building. In a day, a cutter would prepare enough stone to cover 6 square feet. Thus, if you want to mess with cutting stone, it will take you three times longer to build something than if you use rubble.

PHILOSOPHY AND THE STONEMASON

In a society of ever-increasing transience, stonemasonry stands out as a symbol of permanence. It is the exact opposite of ticky-tack. The ancient materials and the ancient technique will still give you a building that will outlast all the competition. If the pyramids had been built of pressure-treated wood, all the information in them would have been lost. If Stonehenge had been made of concrete, it would be in pieces by now. Because stonemasonry, done properly, outlasts all

other building methods, extra care must be taken in deciding what to build with it. It is wise to get it right the first time. In carpentry, if you make a mistake and frame a wall in the wrong place, it's a 2-hour problem to pull the nails and move the wall. In stone construction, the same blunder could cost weeks of effort.

Part of the intention of this book is to steer the reader toward applications of stonemasonry that will give the greatest rewards, while revealing pitfalls along the way. For example, suppose you are going to build a stone house and the stonework is to be exposed on the inside of the house. It takes a long time to get a finished stone wall, and you will want it to be seen, but take a look around the room you are in now. What percentage of the surface of the walls are exposed to the eye? Would you want to spend days fussing with a section of wall and then have it covered with a refrigerator? With all the modern alternatives, a stone wall needs to be seen to justify the extra work required to build it.

A stone wall must also last in order to justify the time spent in construction. Because gathering the materials can be one-third of the work, and fooling with the mortar mix can be another third, the actual rock-stacking part might only be one-third of the whole job. So, if you discover some fancy method that cuts the stone-laying time in half, usually at great cost to the structural integrity, you've actually only saved one-half of one-third, or 16.6 percent—not worth it.

People once built all-stone fireplaces and chimneys. That doesn't mean this method is still a good idea, however. Stone (except soapstone) cannot handle frequent and extreme temperature changes. The rough texture of stone is bad in a chimney because it impedes the flow of gases and encourages deposits of creosote. Fireclay products are a vast improvement over stone for these uses.

If you've never done any masonry, you'll be able to learn stone stacking quickly. If you've done brickwork and block work, your knowledge could be an obstacle. The same principles apply in all types of masonry, but in brick and block work, the emphasis is on uniformity, perfectly straight lines, square and level units, uniform mortar joints, and straight courses. In rubble stonemasonry, on the other hand, all the uniformity is out the window. The units will vary like the wind, with bumps that protrude and dips that don't, corners that undulate and tops that are wavy.

Overall, you can achieve plumb, square, and level with rough stones, but not within each unit. This makes stonemasonry a very different craft than brick and block masonry.

Ten brick masons with identical piles of bricks will build ten nearly identical walls. Ten stonemasons with identical stone heaps, would make ten very different walls, even though they might all use the same technique. There is an artistic aspect to stonework that must be considered. If you aren't artistically inclined, in some way, you probably won't enjoy working with stones. I've worked with carpenters on stone walls, and they would get annoyed with how slow the progress was and how heavy and crooked the materials were. Production carpenters, or 4- x -8 technicians, as I call them, hate stonemasonry. It's hard to fit into the schedule, hard to measure, hard to predict, hard to find someone to do it—it's the antithesis of the 4- x -8-foot sheet. If you don't love beauty, stonemasonry could be a real pain in your plan.

It's safe to say, at this point, that you don't have to be nuts to do stonemasonry, but it helps. It is a favorite medium for artists and curmudgeons, crackpots and visionaries, and hard-nosed individualists. People who have never built a thing in their lives can often build a stunning log cabin, but they will do a miserable plasterboard job. Similarly, first outings in stone are often beautiful. The rough materials are less intimidating to those who have doubts about being able to produce perfect uniformity. Experience sometimes teaches you to cheat and cut corners.

When working with stones, gravity is about all you need to understand. Knowledge of attaching devices and glues is a detriment in working with rubble stone. Rock stacking is so basic, like basket weaving, that your hands probably already know how to do it. Part of the problem is in getting the mind to shut up so the body can get to work. Again, I stress that a person can build a beautiful and lasting structure of stone on the first try.

WHY BUILD WITH STONE?

There are many good reasons to build with stone. The wealthy build with stone because they can afford the very best. They have stone mansions built. Some people are so poor, rocks is all they have. They build stone huts. The rest of us have other reasons.

Beauty

Beauty has become the primary reason to build with stone. In the old days, before ugly architecture was invented, practicality was the main reason for building with stone. Even a mere hundred years ago, many stone houses were whitewashed on the outside to cover the stones. Bricks were more prestigious then, and rubble stonework was not something to show off. There was plenty of beauty in the environment then, so a hand-hewn beam or a rubble-stone wall was a primitive necessity, not a beautiful conversation piece. Old stonemasonry is beautiful by accident. There was no technology to allow an ugly wall. A wall looks bad because the stones don't fit right, and in the old days, if the stones didn't fit right, the wall would fall apart.

Because it is almost always faster and cheaper to build with block, brick, or concrete, stone should have special aesthetic qualities to warrant being used. Aesthetics are certainly a matter of opinion, but beauty has become part of the practicality of stonework. The whole craft is getting squeezed into the role of decoration, and what is the purpose of a decoration save to look nice? Again, a well-built stone wall is automatically nice to behold. The only way to mess it up is to smear it with cement. My method of laying stones greatly diminishes that problem.

Stone walls without mortar almost always look great. For that matter, so does a stack of firewood. There is something pleasing about organized units of natural random components. A stone wall with too much mortar won't be very pleasing to view. If the surface that your eye strikes is more than about one-third cement mix, the wall probably won't look much better than a parged block wall. The stones, and their relationship to each other, are what looks good, not the cement.

A dry-laid wall will look good even if it has 50 percent space because the spaces reveal shadow and the sides of stones—more natural objects. Shadows are as pleasing to the eye as rocks, but cement usually is not. Some stone walls look like large aggregate concrete walls. I like the look of a wall that is built of stones, not cement.

A stone wall that is built well is soothing to the eye because it is at balance, at ease. It doesn't demand attention. If you do it right, you won't have to do anything else to a stone wall in your lifetime. Neither will your children.

Endurance

Endurance is the next most obvious reason to build with stone. A properly built wall will last for centuries. A well-built and well-maintained stone wall can last millennia. Here I must emphasize that a poorly built wall of stone will not last long at all. FIGURE 1-2 shows a 30-year-old stone building in the foreground and a 200-year-old stone house in the background. The new wall is bowed out and crumbling, while the old house looks young. The irony is that it is nearly as hard to build a wall that will last 50 years as it is to build one that will last 1,000. I'm of the opinion that if you're going to go through all the motions, you might as well do the rock-stacking part in a manner that will endure. Your great-grandchildren will thank you.

Strength

Strength is another good reason to build with stone. Most stone has better compressive strength than concrete. The granite piers of the Brooklyn Bridge towers, which support one of the heaviest loads of any structure on earth, are still doing well, while reinforced concrete piers on newer bridges have shown wear.

In residential applications, the great compressive strength of stone is not important. Even at the Brooklyn Bridge, the bottommost stones only carry about 400 psi. The only time the compressive strength of a rock matters around a home is if you crack black walnuts on one with a hammer.

Mass

The strength we normally attribute to a masonry wall is really the result of mass. If you drive a car into a stone wall, the wall doesn't move much because it is heavy, not strong. The tensile strength of a stone wall is also a matter of weight. If you tried to knock a stone out of a wall at the bottom with a sledgehammer, the stone would resist, mostly because of the weight on top. If you took a crack at a stone on the top of a wall with a big hammer, it would move. That part of the wall has less tensile strength because the rock has no weight above to give it great resistance to lateral forces. A stone wall is strong because it is 2 feet thick, not because of the cement powder in it. The Washington Monument doesn't blow over because it is so heavy, not because it is glued to the ground. Gravity has that rock pile glued down tightly.

The strength of a concrete wall is different from the strength of a stone wall. Concrete's strength is more a matter of the cement powder and the way it

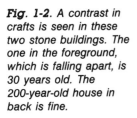

Fig. 1-2. A contrast in crafts is seen in these two stone buildings. The one in the foreground, which is falling apart, is 30 years old. The 200-year-old house in back is fine.

holds on to metal reinforcing. A concrete wall acts as one large stone, and so it can be much thinner than a stone wall without being dangerous. The stones in a stone wall are not glued together. This flexibility is one of the hidden strengths of old-time stonework. The thick walls can handle some motion, while the concrete would crack.

It is hard to build a rubble wall that isn't thick. It takes a certain amount of space to do the balancing act of stonemasonry. A cement block wall will stand free with a mere 8 inches of thickness, but a rubble-stone wall needs at least 18 inches for stability. The old farmhouses where I live have 2-foot-thick walls. These houses are up to eight times heavier than a hollow-core block house of the same size. The funkiness of the stones forces the amazing overkill in the weight of a stone house. If you could find stones shaped exactly like cement blocks, you could build a much lighter stone house. A 40- x -40-foot two-story stone house, with two-foot-thick walls will weigh upwards of 500 tons. Its the tonnage that allows a stone house to stand up to a hurricane, not the cement.

Stone interacts with the environment, just like everything else, but it does it very slowly. For the sake of a few generations, it is safe to say that the wind and rain won't damage stone. Bacteria won't decompose it. Insects won't eat it. Bullets won't go through it. Rats and mice won't chew holes in it. Fire won't burn it away. It is not a bad building material if you don't have to pay for it.

Ecological Awareness

Gathering stones to build a house can actually improve the environment. In very rocky woods or fields, removing stones will allow more growth. More sunlight will strike the soil and more plants will grow to supply more oxygen and clean air. If you have to pick them up anyway, consider putting stones into a rock structure.

Resale Value

If you do a halfway decent job, almost anything you add to your home or yard with stonemasonry will increase the value of the place. People basically love stonework and will pay extra for it. If you have the time and the stones, you can improve a farmstead or even a suburban lot quite a lot without any money input.

If you are lucky enough to have a small stream going through your place, you can build a small dam with stones in a couple of days to add a serene pool and a babbling little waterfall to the scenery. This will get the prospective buyer jumping up and down faster than anything.

In my area, stone houses built before the Civil War often resell for more than the cost of a comparably sized modern home. In a way, they get better with age. Everything around them gets worse and they gain charm. The worth of a historical stone building is even greater. When you build with rocks, you have a good chance of creating a historical building, because it will be around long enough to get some history.

Therapy

Building with rocks is slow, heavy, and very calming. It amazes me how often I find stonework that has been done by a doctor or lawyer or businessperson. People's lives often get very removed from the earth, and stacking rocks will bring them back. It is the opposite of computer screens. If you want to reaffirm your connection with the planet, try holding a 100-pound stone. Then to reaffirm your connection to past and future cultures, build something that will last. Dry-laid walls can be built today that are in no way different than walls built 2,000 years ago.

Necessity

Stonemasonry could be deemed a necessity if you have no money, a lot of rocks, and no shelter. If you live in a remote area where it is hard to get building materials, you can make a good shelter with stone laid in clay.

WHO CAN BUILD WITH STONE?

I am tempted to say that anyone can build with stone, but I must admit, it is not for everyone. You don't need to be big and strong unless you have no ingenuity. Most of the ancient stonemasonry, like the Great Wall of China, was done by small, undernourished people. If you have desire, you don't need much else.

You must have time to get involved with a stone project. If you are in a hurry or under a lot of pressure, it won't work very well. It could require years of work for one person to build a stone house. The largest of the projects in this book took about a month.

COST OF BUILDING WITH STONE

You can lay stonework with no money at all, if the stones are on hand and you aren't using mortar. The cost of food might swell slightly, and you might wreck some clothes, but you can make some significant improvements on a place with zero money. Even if you hire out the job, time—not materials—is the big cost of stonemasonry. Many masons will bid a job by the amount of square feet of exposed stone. It might take a mason one hour, including all aspects of the job, to assemble a square foot of stone wall. If someone else is gathering supplies and mixing cement, he will be able to lay more stone in that hour. His work will probably reflect an hourly wage that is typical of skilled labor in your area.

If you do hire a mason, keep in mind that his skill, which you are paying dearly for, is not driving to get sand or washing out wheelbarrows. It pays to have a tender working with the mason to do the gofer work and the cement mixing. The tending duties are at least half of the job, so if you can get a neighborhood kid to do it for minimum wage or you do that part yourself, obviously the project will be quite a lot cheaper than if the mason has to be his own expensive tender. Mason tending is pretty awful work. It is like being a waitress with 50-pound cups of coffee. Brace yourself.

Even if you have a lot of money, it is unrealistic to think you can call a number in the Yellow Pages and get someone to come and build you a stone house. You might be able to get a crew of brick or block masons that way, but their styles and differing degrees of ability might give a hodgepodge feel to the masonry. It is also unlikely that a single mason will take on a very large job.

If you have money for just a little professional assistance, I suggest you do the following. Hire a masonry crew to lay out your job, pour concrete footings, and set up the lines for the walls. This is the hardest part for an inexperienced builder, and a mistake in layout will carry on through the whole job. If you have any money left after that, hire a stonemason for a few days (tell him your intentions) and be the tender. While you are mixing cement and bringing stone, carefully observe how the materials are being assembled. If you are lucky, the mason will show you what he's doing. Then, after he goes, you do the stonework and try to get a helper to do the tending. If you can only hire a stonemason for one day, have him help gather stones. Inexperienced stone gathering will set a more disastrous precedent for the rest of the job than even a bad layout. It takes experience to know what to look for in a rock.

WHEN TO BUILD WITH STONE

Using stone as a building material makes for some tricky timing. It is difficult to predict how long a project will take, and when it is related to another project it can cause some problems. Suppose you build a stone foundation for your new house, and it takes twice as long as you predicted, which is normal. If the runover time takes you into freezing weather, you will have to stop production until spring. If you already had carpenters lined up to frame the house, the slow foundation can throw the whole schedule haywire.

The Weather

If you are building where it freezes in winter, the weather can be a real thorn in the side of masonry. If the mortar in the wall freezes before its water has had a chance to become part of the new material, it won't ever get hard. Days later, you will be able to crumble the damaged mortar between your fingers.

Frozen sand is also no fun to work with. Even if you manage to thaw it out, it will be so cold that the mortar it makes will need days to set.

Rain is as much a problem as freezing. A hard-driving summer cloudburst will wash the mortar out of a fresh-laid wall and leave the stone faces covered with runny cement smear. While you run around frantically looking for a plastic tarp to cover the mess, the rain is busy ruining the bags of cement you left uncovered. The sand heap starts to wash down the hill as you watch the plastic tarp blow into your neighbor's garden. You'll spend the next day trying to clean the mess off your once pretty rocks with a wire brush.

The worst part about the rain is its unpredictability. If you are a mason and it looks like rain when you get up in the morning, do you make mortar or not? Often, it will be raining in the morning and stop at noon. Do you start a short day's work? Many days you can't tell what is coming. If you live in California, please disregard this critique of the weather.

Too warm can be as bad as too cold or too wet when you are working with cement mortar. On a sunny, hot day, the water in the mortar will evaporate out of the mix before it has had a chance to become part of the new material. If the sand or the water is hot, it will speed up the reaction and cause a weak, premature set in the mortar. Mortar, like concrete, needs a long curing period to achieve its maximum strength.

Darkness

Unlike the weather, darkness is easy to predict. It doesn't really sneak up on us. Stonemasonry is so hard to predict, however, that darkness can become a problem. Suppose darkness falls at 6 P.M. At 3:30 P.M. you decide to mix another batch of mortar. You accidentally get too much water in the mix, so you have to add more sand and cement to get it right. Now you've got a giant batch of mortar, and rather than throw away the extra, you work till 5:30 to use it up. The sun is setting as you wash out the wheelbarrow. If you do like most masons, you've left the rough mortar squish out between stones as you lay, with the intention of dressing down the mortar a little later, when it has stiffened. It is a two-part procedure, as you will learn more about later. So you rush through the dressing part of the job to beat the darkness.

This situation happens so often to stonemasons, and it is such a pity to have the finished part of the wall be the most hastily done. The finish work should be the least hurried. If you put the dressing-down part of the job off until the next day, the mortar will be hard enough to make smooth scraping of the joint almost impossible. FIGURE 1-3 shows a wall that was hastily finish-pointed. The shame of it is, the house shown in the illustration is nearly 200 years old and well built. Now it will look pretty shabby for another 50 to 100 years.

Beating the Weather

You can't lay stones in mortar if it is going to freeze within 24 hours. You can't lay stones if it looks like it is going to rain. You can't do it if it is going to be too hot. You can't work in the dark or when you are too fatigued. If you play it by the book, darn few days in Pennsylvania are fit for stonework. They are probably the same few days you are allowed to do exterior latex painting, according to the directions on the paint can. You must cheat if you want to get the house painted or the stone wall built.

There are many ways to avoid the negative effects of the weather and create more suitable circumstances for stonemasonry. The rain is less intimidating if you have a tarp all ready for emergencies. I keep

Fig. 1-3. A sloppy pointing job spoils the look of a Civil War house.

cement off the ground on a pallet near the job, with a cover that stays on all the time. This is the only way I could figure to avoid getting caught with spoiled bags of cement. In southern California, in the summer, I guess I would leave uncovered cement overnight without worry, but in the Northeast, it is not worth the risk, no matter how clear it looks when you quit for the day. In fact, if there is no rain in sight, an uncovered bag of cement will bring rain.

An occasional mist of cool water from a garden hose will allow you to lay stones in the hot sun. The water evaporating off the wall will keep it cool, and it will keep the mortar from drying too fast. A cover of straw over a newly laid section of wall could keep it from freezing on a cold night. Improved timing and a spotlight will decrease the hassle of darkness.

Still, the weather will mess with you and cause setbacks in masonry. In fact, needing to erect shading screens or heating tents, or even covering a wall with straw are setbacks of their own. It takes extra time to protect your job from the climate.

Work Inside

Whenever possible, keep your project under roof, where you will have more control of the weather. Most of the projects that illustrate this book were done under roof. The Trombe wall shown in FIG. 14-1 was built with the roof structure already up on temporary posts. The wall then was built up to meet the rafters and the posts were removed. It took a day to set up the temporary roof support, but it enabled two men to work without weather hassle on a two-month project.

A hard rain on a freshly pointed stone wall is something to be avoided. The two-step pointing procedure I use will greatly reduce the chances of rain ruining your job. This procedure is explained in more depth throughout the book.

Work without Cement

Another great way to gain freedom from the confines of the weather is to build walls without mortar. Dry-laid walls can be built in the pouring rain and freezing cold. I have built walls in 110°F heat and −10°F cold. If the wall you want doesn't need cement mortar, don't use it. Not only will you gain immunity from the weather, but the work will be much faster and cheaper. It also will look better.

PREDICTING TIME

It is nice to have some idea how long a project will take. If you have never worked with the materials before, time is very hard to predict. I've never heard of a stone project that went faster than expected. If you get an intuitive feeling and multiply that by three, you'll be close to predicting how long it will take.

Beginners tend to wildly overestimate how much will get done. I think this is the correct attitude. Delusions of grandeur really tend to get a project going. If you knew beforehand how hard something was going to be, you would never do any of the great stuff like mountain climbing and hole digging.

In a more scientific vein, the few times I have actually kept track on a stone job, it took me roughly 10 hours to assemble 1 ton of materials. This was on a wall with mortar and a snazzy finish face, and includes every aspect of the job, like driving for cement and washing off the tools. How far will 1 ton of materials go in stonemasonry? I figure the weight of a solid stone wall to be 150 pounds/cubic feet (lbs/cu ft). At this rate, 1 ton will contain 13.3 cu ft of masonry. If you are building a wall 1 foot thick, 1 ton of materials (stone and mortar) will cover 13.3 sq ft of the wall. A 6-inch facing will cover 26 sq ft per ton.

A cubic yard of stonemasonry or concrete weighs about 2 tons. Mortar is lighter than most stone, so the greater the percentage of mortar, the lighter the wall. Dry-laid rubble walls weigh much less than 150 lbs/cu ft because of all the empty space.

Another estimate of time on a stone job might be one hour for each square foot of face stone, including all parts of the job. With a helper, this amount could easily double.

To figure quantities of sand and cement in a wall, assume the wall might be one-third mortar, and the mortar might be one-third cement. Each cubic foot of laid rubble might require 10 to 15 pounds of cement, and nearly 50 lbs of mortar. Don't let these or other figures be intimidating because there is much leeway in rubble masonry. If you have doubts about predicting time or quantity, just keep track of a small section of work and see how long it takes and what it used. Then you will have a pretty good idea about the rest of the job.

You need extra of all the materials. It is very unlikely that when you use up all the stones the wall will be done, like a big puzzle. Make this an easy puzzle for yourself and have extra pieces. It is especially silly to run out of sand. Where I live, the trip to the sand yard is much more expensive than the sand itself. If you need much more than a ton, you should get it delivered. If you need slightly less than a ton, you should get a ton. Where I buy sand, 1/2 ton cost more than 1 ton. It is billed at $1.00/100 lbs or about $9.00/ton. I can always use the extra for the garden or kid's sandbox. A ton of sand, incidentally, makes a big sandbox, and could be the best toy you'll ever spend $10.00 on.

On any large job, like a garage-sized building, you will need enough sand to have it delivered in a dump truck. Don't think you save money by getting the sand in a pickup truck. You have to shovel it out of the truck when you get home. For a stone house, having sand delivered in a dump truck can save you 100 tons of shoveling.

Again, as a rough rule of thumb, I'd give yourself 1 hour for every square foot of stonework in your project. For laid walls that aren't seen and demand no face finishing, halve the time. Stone foundations undergrade and dry-laid walls are roughly three times faster to build than finish-pointed stone walls. I've heard that a good stonemason can build 50 square feet of wall face in one day. I suspect this is with a tender keeping the materials coming. I bet it is during ideal weather on one of the longest days of the year. I suspect it is work at ground level on a very straightforward wall without openings. Quite possibly the stonemason who started this rumor was bragging 10 to 20 percent. In conclusion, don't be disappointed if you only knock off 10 square feet of stonework in a day.

I don't want to dampen your spirit for the outrageous, the excellent, or the unique. If you are going to climb a big mountain, however, you should be prepared to climb a mountain. Building a real stone house as a solo venture is in the league of climbing in the Himalayas or sailing around the world—wonderfully impractical.

PRACTICAL APPLICATIONS

Fireplace inserts imply the impracticality of the masonry fireplace they are inserted into. They make folly of the expensive work of the mason in an attempt to turn the fireplace into an efficient heater. Many people have blocked off the fireplace and put in a wood-

stove. I no longer think it makes sense to build masonry fireplaces, unless you have money to burn, or want a special atmosphere, like at a ski lodge.

Outside on a Hill

I think the most practical use of stonemasonry, especially for the novice, is for dry-laid retaining walls. The work is fast and forgiving. The rocks don't need to be great, and there is no mortar or related problems. The weather can't ruin the work; no tools are needed; and no money is invested. The work almost always looks great because it is nearly impossible to make an ugly dry-laid wall. The rear side of a wall that is banked with earth can run ragged, or be very non-uniform. A retaining wall serves a practical function in creating level areas on sloped land. If you keep the wall low (3 feet), you won't have to worry about great structural integrity or fooling with scaffolding. Dry walls require very little manual dexterity. The work is great in the winter. You can do it with mittens and there will be no flies. The work is heavy and fast and will keep you warm.

Earth-sheltered houses are gaining popularity, as well they should. Stone retaining walls go very well with these designs. FIGURE 1-4 shows a dry-laid retaining wall that keeps the dirt against an earth-bermed solar house. The wall was built in 2 days by a novice, with almost no money during bad weather. It looks great, helps save heat in the house, and creates a level area in the yard.

Near a Pile of Rocks

The most practical of stone projects will become very impractical if you aren't near a good source of free or very cheap stones. If you happen to own land that has an old stone building or ruins on it, it would be most convenient for you to rebuild right next to the ruins using that stone. On small projects, the proximity of the stones is less significant.

A stone house, like the one shown in FIG. 1-5, probably has more than 100 pickup truck loads of rocks. If the rocks were 10 miles away, you'd be doing 2,000 miles of hard driving during the job. The house shown in FIG. 1-5, incidentally, is well over 200 years old, and it is not even a very well built rubble house. There are many poorly shaped stones, and the window and door openings are spanned with wooden lintels. You could build a house this good with a lot of time.

If you ever get involved in a 100-ton job, remember that it makes sense to lift each stone the minimum number of times. Suppose you are bringing stone to a building site in a pickup truck, but you can't pull the truck right up to the wall because a tree is in the way. You take the rocks out of the truck and put them in a wheelbarrow and push them to the wall. On a 100-ton job, you've just doubled your lifting by not cutting down the tree. Suppose you dump the stones out of the wheelbarrow onto the ground and then pick them up to put in the wall. You will now do three times more lifting than if you could pull the truck right up to the wall and work the rocks directly into the wall from the truck.

Fig. 1-4. A dry-laid retaining wall built by a novice.

Fig. 1-5. A 200-year-old house in Gettysburg.

When you're talking about umpteen tons, it is worth a few moments' thought and a day's work to avoid an extra lifting of each stone. It is not unusual to lift a stone ten times before it is finally at rest in a wall.

There is nothing romantic about lifting stone. If you want big biceps, work out with weights. Stone lifting will only give you tough hands and a sore back, so watch each time you lift them. The ideal is to have a dump truck dump the pile right where you need it. On smaller jobs, it is much more fun to go out and find stones in the woods, on the mountains, and in the streams. Each rock becomes special.

Inside in the Sun

Another practical place to use stone work is inside a house in the path of direct winter sunshine through glass. If you have, or are planning, a solar house without much mass, the interior temperature will rise dramatically when sunlight penetrates the glass, and the temperature will drop as fast when the sun sets. A wall of masonry will absorb some of the extra heat to release after the sun has set.

Chapter 14 describes what I think is about the most practical use of stonework that is possible. I have designed a wall that absorbs heat from the sun and backs up a wood stove to store its extra heat, too. The wall encloses the stove pipe and supports the ceiling or roof of the house. It also encloses a large planter and functions as a room divider. If the rocks for a project like this are removed from your soil to benefit a garden, you've done the most with the least effort.

WHAT IS A STONE WALL?

A stone wall is a lot like a large, three-dimensional puzzle. The dimensions of the finished puzzle are whatever the dimensions of your desired wall are. If your wall is freestanding (not against a block wall), built of rubble stone, more than a few feet tall, it will need to be at least 18 inches thick. It requires this much space to do the balancing act. Walls 24 inches thick are easier to fit, but require 25 percent more work. Walls a mere 12 inches thick are a possibility if built against a plumb temporary brace and if they don't support anything—such as a fire wall behind a wood stove. Even thinner walls could be made for low planter walls and partitions, especially if one side of

the wall can be *parged,* or completely smeared with a layer of sticky mortar. Generally, a real rubble stone wall is at least 18 inches thick.

For a stone wall, I recommend making everything as easy as possible. Use pieces that are easy to visualize together. Have plenty of extra pieces to increase your options, and if a piece won't fit quite right, break it so it will. This isn't cheating; it is part of the deal with stonework.

A stone wall must stand up and be plumb to pass the test, and there is no way you can cheat. The whole magic of rubble masonry is that heavy, flat objects are stacked on top of each other so that the weight is always falling straight down. Suppose you needed a wall 18 inches thick, 10 feet tall, and 20 feet long. The best stability you could hope to have with a wall this size, would be a single large stone with a flat bottom that sat on a flat footing. Yet when you envision this stone, it seems rather precarious—like a huge domino standing on edge. A dinosaur could perhaps push the thing over. To us humans, however, it is very stable. Nothing in our world can normally affect the stability of the giant stone on edge. If the stone had a sloped bottom, and sat at an angle, just shy of the tipping-over point, then obviously our bodily efforts could knock the wall over.

The strength of a stone wall is its tremendous weight, always trying to fall straight down on another stone. No magic, no glue, no connecting rods . . . just gravitational force.

I get a chuckle when I see people trying to build a laid stone wall with reinforcing metal. The rebar actually becomes the weak link in the wall. A vertical joint is imposed by the metal and doesn't allow legitimate overlapping and interlocking of stones. When a stone wall with vertical rebar falls apart, it will happen right where the rebar is. The stones will crumble in a heap, and the reinforcing metal will remain poking skyward, all covered with the mortar that adheres fabulously to the metal. The metal is wonderful in concrete and ferroconcrete, but worse than ridiculous in a laid stone wall.

I have seen rebar used horizontally in stone walls as well. I wish someone could tell me what the purpose is. Stonemasonry works. Time has proven this without a doubt. To build a successful stone wall, you won't need anything that wasn't available 1,000 years ago.

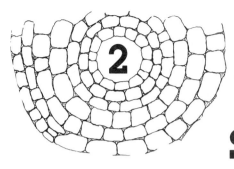

2

STACKING AND GATHERING STONES

THIS chapter title might sound like two different subjects. It is. The paradox is that you cannot instruct what type of stone to gather if the reader doesn't already know something about stacking them. To prepare you for gathering stone, I have included some very basic experiments with stacking.

MASONRY IN MINIATURE

You can learn a lot about masonry and laying stones using miniature units. The same basic laws, rules, and advice hold true in miniature, so you can find out about them without busting your hands and back and truck. A sugar cube is a convenient imaginary masonry unit. Let's build a wall, at least mentally, of sugar cubes. Each cube will represent a scale cubic foot of stone, and the wall will be 2 feet thick, 10 feet long, and 8 feet tall—a total of 160 cubic feet.

Foundation. Before you begin the pretend wall, you will need to know where to build it. The wall needs a solid foundation. You wouldn't want to build it on a pillow, because the pillow will give too easily and the wall will tumble. You wouldn't want to build it on a block of ice because the ice would melt and the wall would tip over. You couldn't build it in a sandbox because the sand shifts too readily and the wall would fail. If you started the wall on a wobbly legged card table, it would get bumped and the wall would tumble. You need to build on a surface that won't move, change, or wash away. For the purpose of this experiment, a sturdy tabletop will do nicely as a foundation.

First Course. This wall of sugar cubes is to be built without mortar, that is, dry-laid. It will be built in horizontal rows, or *courses*. There will be eight courses, each 1 scale foot in height. Draw a straight line along the tabletop and align ten cubes on the line. Without pushing anything out of line, put ten more cubes behind the first row. Make sure the ends match up. You now have a scale wall, 10 feet long, 2 feet thick, and 1 foot high. Your wall has two faces, two ends, and a top (FIG. 2-1). If you kept stacking the cubes in the exact same matter, until they were eight layers high, the wall could cleave in two, or split at one of the many continuous vertical fissures, or joints. You must prevent those vertical fissures from occurring.

Second Course. Before beginning the second course, you must cut two cubes in half, so you'll have four half-cubes. Begin the second course with half-cubes, then lay nine full cubes and end with another half-cube. Starting with the short unit has allowed the staggering of vertical joints in the wall. Lay the opposite face of that course the same way. FIGURE 2-2 shows three views of the wall at that point. The vertical joints are crossed, but there are two new problems: the wall is still in two separate sides, and the half-cubes that began the second course might tend to fall over sideways, out of the wall.

Third Course. Before you start the third course, you need some *tie stones*. These are rocks that span the thickness of the wall, tying together the opposite faces of the wall. You will have to glue two cubes together to create tie stones. In masonry, you would find bigger stones instead of gluing two together, but for the sake of this analogy, you will have to pretend that the glued cubes are single, large units. Start the third course at the ends, with the two tie stones crossing the internal joint of the wall. Make sure they are set plumb with the existing wall, by sight. FIGURE 2-3 shows how the tie stone secures the rather shaky half-cube below it. The two sides of the wall are now less likely to cleave in two.

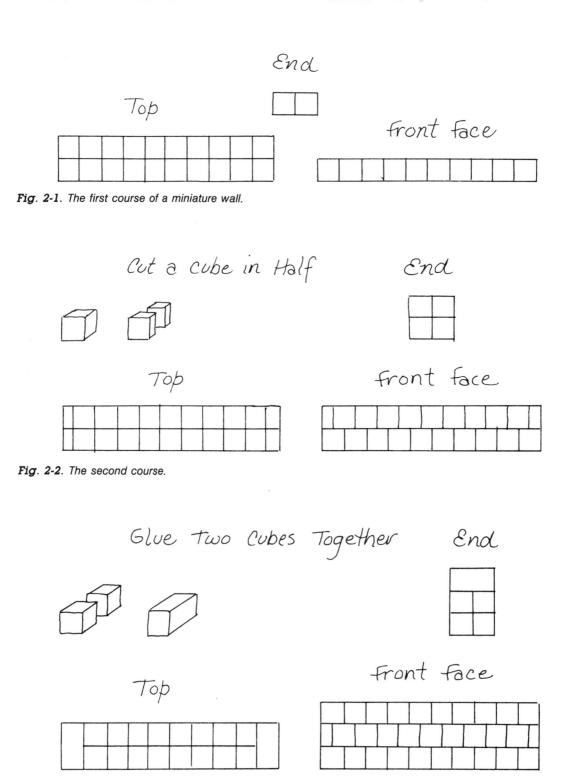

End

Top

front face

Fig. 2-1. The first course of a miniature wall.

Cut a cube in Half

End

Top

front face

Fig. 2-2. The second course.

Glue Two Cubes Together

End

Top

Front face

Fig. 2-3. The third course.

16

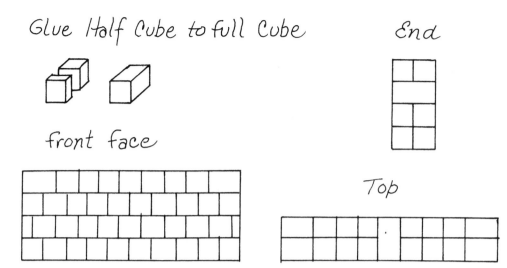

Glue Half Cube to full Cube

End

front face

Top

Fig. 2-4. *The fourth course.*

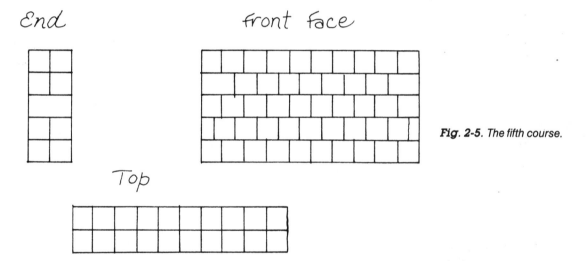

End

front face

Top

Fig. 2-5. *The fifth course.*

With the new end pieces in place, add the other 16 cubes of that course, working from the ends to the center. If the last two cubes don't quite fit, you must either stretch out the course, or cut some of the block. If there is extra space when the last two cubes are installed, you must either squeeze the course in or add some sugar to the crack.

Look at the wall now. What are its weaknesses and how can you correct them? Probably the wall is starting to lean one way. If you weren't going any higher, it might not matter, but to get the scale 8 feet, you need to adjust things to keep a plumb wall with a level top. Assuming the tabletop is level, you can

check the sides of your wall for plumb and the top for level by eyeballing across the side or the top of a square object held against the table. If you stand a book up on end on a level table, the side of the book will be plumb and the top will be level. Using one eye, get on one side of the book or whatever square object you sight across, and align that edge with the top of your wall. Do they coincide, or does your wall angle away?

If your wall leans, adjust it by adding or subtracting sugar and then sight the ends and top again. Primary to the theory I'm trying to bring out is to pay attention to the particular circumstances. There is no script for rubble stonework. If you happened to have a weighted string hanging from a beam, you could sight across that to check your wall for plumb. If the wall isn't leaning at all, you can keep building.

Fourth Course. To build a stable wall, you now need a new size of block. This time, glue a half-cube to a full cube to get a block a scale 18 inches long × 12 inches high × 12 inches wide. You might start the fourth course with a pair of these blocks at either end, with the long side as the face (FIG. 2-4). Then lay the remaining 14 cubes of that course. A tie stone (double length) in the center of this course would be a nice gesture (SEE FIG. 2-4, TOP VIEW).

I use this sugar cube analogy not because the cube is an ideal shape for masonry, but because it is an easy metaphor to grasp, and you might have some around to play with. Dominoes are actually the premier tabletop masonry unit.

Notice the wall face at this point. All the joints have been crossed. Use tie stones to bridge the internal crack occasionally.

The Fifth Course. The fifth layer of sugar blocks could be done as the first course was, and the vertical face joints would remain staggered from the course below (FIG. 2-5).

Sixth Course. A repeat of the second course keeps things well integrated on the sixth course. (SEE FIG. 2-6.)

Seventh Course. FIGURE 2-7 shows the seventh course, which is the same arrangement as the third layer.

Eighth Course. The top course, if possible, should be all tie stones, or capstones. They will do the most to knit the two faces of the wall together, and heavy units on top will be less likely to get bumped off the wall. Because the seventh course begins with tie stones at the ends, the cap, or top, must be done differently at the ends, or the joints between the end capstones would align and make a weak spot. Therefore, I have begun that top course with the 1 1/2-cube units (FIG. 2-8) to create a stagger with the cracks below.

It is a good idea to get the top of the wall level. You can eyeball level using a known right angle (90 degrees) against a known plumb line, sighting across the horizontal plane, and attempting to bring the wall top into parallel alignment. If you hung a yo-yo very still and held a box of cereal along the string (without touching the string), the top of the box would be level and you could sight across it with one eye to see if the wall in question is parallel, or also level. If it isn't, you can make adjustments by adding or subtracting material. No mason would use yo-yos and cereal boxes as plumb bobs and levels; I mention them here to remove any mystique these allusive conditions called plumb, level, and square might have. *Plumb*

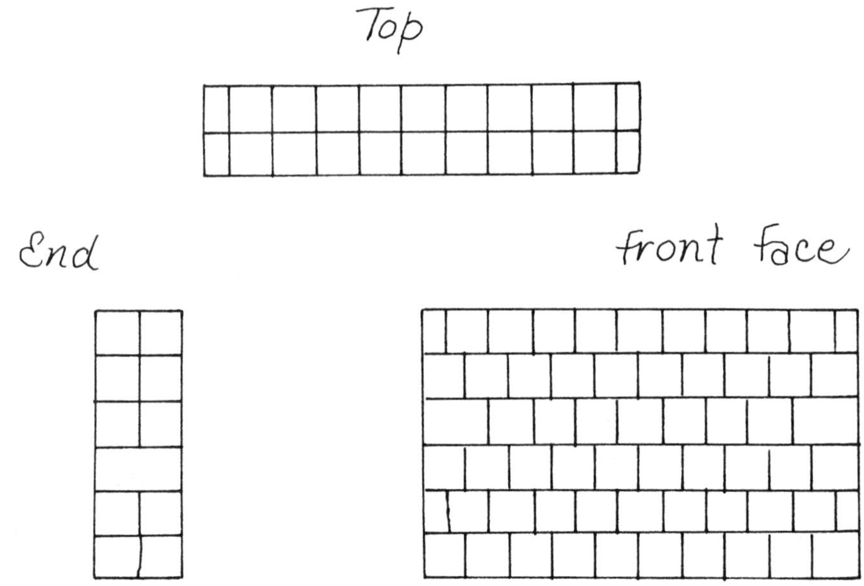

Top

Fig. 2-6. The sixth course.

End

Front face

18

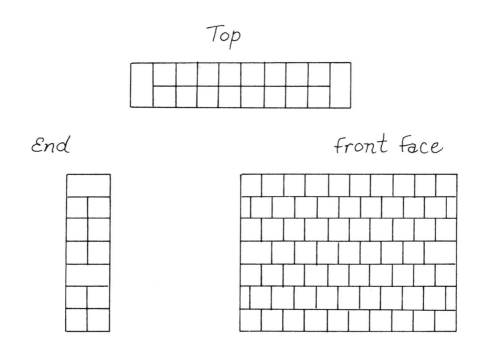

Fig. 2-7. *The seventh course.*

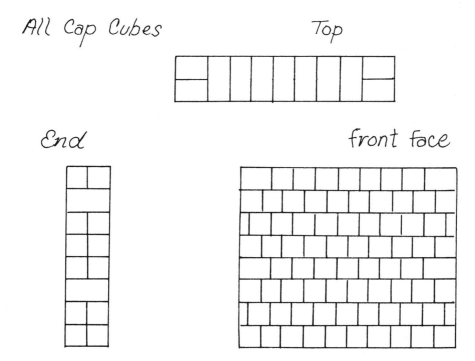

Fig. 2-8. *The eighth course.*

19

is the direction gravity pulls. *Level* is 90 degrees from that. Any number of primitive devices found around the house can be used as references to check these conditions.

If this project had been made of stone, naturally the shapes would have been much different, but it still would have been built the same way: in courses working from the ends to the middle. Combinations of smaller components would probably be used in stonework to bring things up in taller courses. In true scale, this wall would have taken the beginner 16 days, and would weigh 12 tons. This sugar-cube experiment was meant to give you a taste of the possible procedure of masonry. If it was confusing, intimidating, or just too stupid to go along with, try this next experiment and you'll learn how to do rubble stonemasonry.

BUILD A MINIATURE STONE WALL

If you gather a bucket or two of small stones, not bigger than a golf ball, and try to assemble them without glue or mortar, into a wall about 3 inches thick and 12 inches tall, you will learn more about stacking stones than a whole book could tell you. The rules will unfold before your eyes and hands, and the project will be no more tedious than a puzzle, and probably fun. You can build this miniature wall on a tabletop and be very comfortable, whereas learning with real scale is slow and painful.

Probably the first thing you will learn is that roundish stones are hard to stack, and long, flat stones are easy. Make this puzzle easy and toss out the round stones. If you can find shapes that are midway between a domino and a marble, the easiest and hardest shapes to stack, you'll have something like typical rubble. If you spend a few hours fooling around with the pebbles, you'll save yourself many hours later, if you plan to do much rubble stonework. Don't worry about lines and levels with the miniature stone wall. At this point, that stuff would be a distraction to the educational process. The wall you try to build will teach you about plumb and level, and you will get a more natural feel for those conditions, which will help a lot even when you're using lines and levels.

I tried a little wall on top of a crude picnic table in my yard. I really enjoyed the process. It was so fast being a giant, I built a wall in less than a day. The same principle kept popping up as I stacked my pebbles, same as a real wall. FIGURE 2-9 shows the beginning of the miniature wall. I used the edge of the board as a straight line and followed it. The wall is 3 1/2 inches thick.

FIGURE 2-10 shows a closer view of the top of the miniature wall with the next course of stones in place. Notice the tiny gravel in the middle of the wall. This size of stone is useful for raising larger stones into a more stable position, and for filling spaces between

Fig. 2-9. The start of a miniature stone wall.

Fig. 2-10. *The top of the wall shows tiny fill stones.*

larger stones. Notice how the wall looks like a real, dry-laid stone wall. It is. The only thing phony about it is that I was able to use more big stones than I normally would in true scale. If this 3 1/2-inch-thick wall represents a 24-inch-thick wall, there are quite a few 200-pound stones.

The rough shapes are brought up together in a row, but the ends stay higher. The new course begins at the ends, and the middle does not get above that height. If you honestly try to do this yourself as a puzzle, fitting the pieces to make a rectangular whole, the old methods of laying stone will reveal themselves. Stones will spill out of the wall or it will split in two, and you will intuitively know how to prevent it.

Take a close look at FIG. 2-11. The wall is another course taller. See along the top of the wall, that one side is built up, and the opposite face is not. A wall is brought up one side at a time like this, with frequent tie stones to connect sides. The only way the tie stones can work is if the separate sides are continually brought up to the same level. In the illustration, the low face will be brought up to the same height as the high side, and the new course will have stones that bridge the crack. The end of the wall in the foreground is like a cross section of the interior. See how the tie stones hold the wall together.

Faces have been laid on both sides of the wall, with a considerable space left in the interior of the wall. This space will be filled with carefully placed junk stone filler, taking care not to get the fill above the height of the faces, which would prevent a flat tie stone from sitting well on the following course.

The wall is still quite stable and could go higher as long as everything is balanced. I would need to use a level or some sight lines to go a lot higher because the wall would probably start to lean, thereby losing stability. By trying a miniature wall, if nothing else, you will learn what shape of stones to look for when you build a real wall. Knowing what shapes to look for in a rubble stone pile is probably the most skilled part of rubble stonemasonry, and the single

Fig. 2-11. *The new course is unfinished on one side.*

most important thing to know in building a successful wall.

This and the previous exercise have prepared you for more information which, hopefully, you will grasp more readily because of the experience. More detail on shapes and how to assemble them is given in Chapter 4.

GATHERING STONES

Most people I have worked with have confessed to having no idea how much stone they needed for a project, and most of them were surprised to see how small a ton is. A chunk of granite a mere 24 × 24 × 36 inches weighs a ton. FIGURE 2-12 shows some common materials in masonry and the amount of space one ton of them would take up. Don't be fooled into thinking you can fill a pickup truck with stone. There's room for five tons of stone on a half-ton truck, but that doesn't mean the truck can haul it.

In addition to tons, it is a good idea for you to know something about cubic feet. Many things in masonry are referred to with that measurement. FIGURE 2-13 shows various forms of a cubic foot. Because there are 1,728 cubic inches in a cubic foot, any rectangular shape that has length, width, and depth measurements which, multiplied together, equal 1728 contains a cubic foot. The 4- × -8- × -54-inch shape in the illustration is an example.

A cubic yard is also a common reference quantity in masonry. Concrete is sold by the yard. A cubic yard of concrete, incidentally, weighs over 2 tons. Masonry cement is sold in cubic-foot bags. Hydrated lime is often sold in 50-pound bags, which contains slightly more than a cubic foot. It is handy to know that 1 1/2 buckets that hold 5 gallons equal a cubic foot. Most masons can hardly function without lots of 5-gallon buckets, but few of them use them as a measuring device. If you were instructed to mix cement and sand in a 1:3 ratio, you could use a full bag of cement and 3 cubic feet of sand, or 4 1/2 buckets. This is much more accurate than the usual method of counting shovels of both ingredients. Make sure your buckets hold 5 gallons before you use them this way.

Weighing the Job

A ton is a convenient measuring unit in stonework. That's about all you can get on a pickup truck

or a trailer pulled by a car, and those are by far the likeliest vehicles at your disposal for stone gathering. People wisely won't lend you their nice new truck to fetch rocks. Using a car to gather stone is pretty awful, although it might be o.k. for a small job like a hearth or a column to hold up the mailbox. For anything large in stonemasonry, you'll need a truck or a trailer, unless you have stones delivered or happen to have rocks within a wheelbarrow's drive from the job site.

Suppose you want to build a wall behind a wood stove to protect a wooden wall from the heat, to store some extra Btus, and to add something of the flavor of a stone fireplace to your stove. If the masonry heat wall is to be 1 foot thick, 7 feet tall, and 7 feet long, the total cubic feet of the project is 1 × 7 × 7 = 49 cubic feet. Because there are about 13.3 cubic feet of masonry in a ton, divide your total cubic feet by 13.3 to find the tons you need. In this case, 49 ÷ 13.3 = 3.68 tons. It is nice to have extra stones to choose through, so for this job, I would gather four pickup loads of rubble.

You can get a feel for what a ton is on a truck by getting a ton of weighed gravel at a quarry and seeing how the truck handles and how far down the springs go. It is more accurate to visualize the amount of space the ton will require. If your stones were all 12 × 12 × 12 inches, 13.3 of them would be a load. If your stones were roughly 6 × 12 × 12 inches, 26 of them would make a ton. A common stone that size would weigh 75 pounds.

Since the rubble stone you will be gathering will probably be shaped very little like a cube, the load will have a lot of spaces, and therefore will take up more than 13.3 cubic feet of your truck bed. A random heap of big rocks dumped from a truck is about 40 percent spaces. If the bed of the truck is 4 × 8 feet, you could put about one full layer of stones that averaged 6 inches high. If you managed to fill all the spaces on a 4- × -8-foot bed 6 inches deep with rocks, you'd have 2,400 pounds—an overload for most pickups. Don't overload, and be careful. A full load handles poorly. Engines are more apt to overheat, tires are more likely to pop, and brakes are more likely to fail when the truck is loaded.

Pocket calculators make it incredibly easy for you to estimate the supplies you need. If you were planning a stone wall 1 1/2 feet thick by 10 feet long by

8 feet tall, and wanted to know how many stones to get, turn all the dimensions to inches and multiply to get the total cubic inches in the wall. In this example, 18 × 120 × 96 = 207,360 cubic inches. Divide this by 1,728 to get the total cubic feet. 207,360 ÷ 1728 = 120. Multiply the cubic feet by 150 to get the pounds of stone. 120 × 150 = 18,000. Divide the pounds by 2,000 to get the tons. 18,000 ÷ 2000 = 9 tons. If the wall is one-third mortar, which is typical in a laid rubble wall, you would need 6 tons of stone for this job. I would get 7 loads to be sure.

Finding Building Rocks

Like anything else that can't be bought at a shopping mall, you have to put out feelers to locate building stone. Here are the common methods I use to find stone for masonry:

☐ Buy rubble from a limestone quarry. If the quarry isn't nearby, this method would be prohibitively expensive.
☐ Get stone off someone's property with their permission. This is best done when the stones are in the owner's way and you are actually doing a favor to remove them.
☐ Buy stone from state or national lands. I frequently get rock-gathering permits at my state forester's office. In Pennsylvania, they charge $4.00 per ton and it is done on the honor system. This is how I get colorful mountain sandstone with lichens and moss for special jobs.
☐ Keep your eye peeled for old buildings, fence rows, foundations, and other ruins and ask the owners if they will part with the rocks. (They usually won't.) For $5.00 per ton, some of them will let go of an old stone building. It is not worth much more than that if you have to haul it and clean up the site.
☐ Get stones from farm fields in the winter. Most farmers will be glad to have someone clean their fields of big rocks, if it doesn't interfere with agriculture. Most of the old rubble-stone farmhouses in my area were built of stone that had to be removed from tilled fields anyway.
☐ Look for creek stones. Some mountain streams are completely choked with building-sized stones,

and often, the prettiest ones are underwater. Get permission from the owners, and only take rocks that aren't playing a crucial role in antierrosion. If all the stones are taken from a streambed, the water will run faster and cause more damage.

If you are using a vehicle to gather stone, naturally you'll want to get as close to the rocks as possible. It might be tempting to grab the stones right along the sides of steeply banked roads. Don't remove these stones; they perform a valuable service in keeping the road intact.

There are many areas of the Eastern United States that are so rocky people would think you were nuts if you asked where you could buy some. Most Appalachian areas are strewn with stone. The Ozarks are so rocky, it is hard to get a pitchfork in the ground. The Rocky Mountains are aptly named. In more civilized, agricultural areas, stone walls can still be seen between cornfields and property lines. Sometimes, these walls get in the way of progress and can be had for the taking. One of the underground stone houses in Chapter 14 was built from such a wall.

A rock hound will find rocks. If you're driving to work in the city and you notice a road crew tearing up an old stone road, by all means stop and ask them what is to become of the pavers (stones). Don't be shocked if they are destined for the city dump. I've found fine cut granite ashlar and arch stones in a dump. If you can intercept a dump truck load of building stones en route to the dump, and have them dropped off at your building site instead, and get paid for the dumping privilege, that would be the ultimate score in stone gathering. (It will never happen.)

After fires in old buildings, there is often stone for the taking. It is rather morbid scrounging materials from a disaster. Sometimes you can bid on the materials of a building slated for demolition. I have obtained stone foundations from wooden houses that carpenter friends take apart.

FIGURE 2-14 shows a dilapidated outbuilding on an old homestead. The owner was eventually going to bulldoze the little stone building, so he was happy to take $5.00 per ton for the stone. I was able to get my truck right up to it. Notice the walls of the building are covered with a thick lime stucco. This plaster had come off at the corners, revealing some nice corner stones. Old stone buildings weren't built to show off, so they were often weatherproofed with lime. The old lime comes off easily without staining the stones, so this type of ruin is worth bothering with. Stones that have been smeared with a portland cement mortar are not worth bothering with, because the cement, which is nearly impossible to get off, is usually on the best show side of the rock.

Types of Stone to Gather

Now that you have some idea how much stone

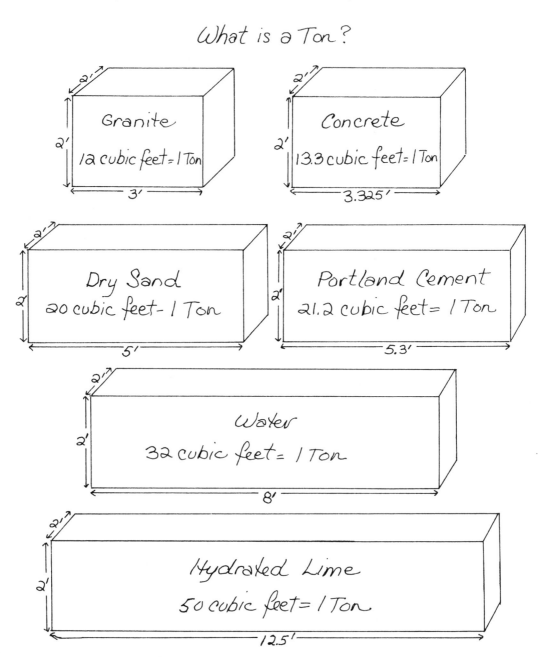

What is a Ton?

Granite
2'
2'
3'
12 cubic feet = 1 Ton

Concrete
2'
2'
3.325'
13.3 cubic feet = 1 Ton

Dry Sand
2'
2'
5'
20 cubic feet - 1 Ton

Portland Cement
2'
2'
5.3'
21.2 cubic feet = 1 Ton

Water
2'
2'
8'
32 cubic feet = 1 Ton

Hydrated Lime
2'
2'
12.5'
50 cubic feet = 1 Ton

Fig. 2-12. *Different size tons (all figures approximate).*

you will need, and where to look for it, you need to know what to look for. Some stone is clearly not worth gathering. If you collect chunky stones like the ones in FIG. 2-15, you can build a wall, but the stones will be difficult to fit together and the joints will be hard to span. This retaining wall has split apart at a line

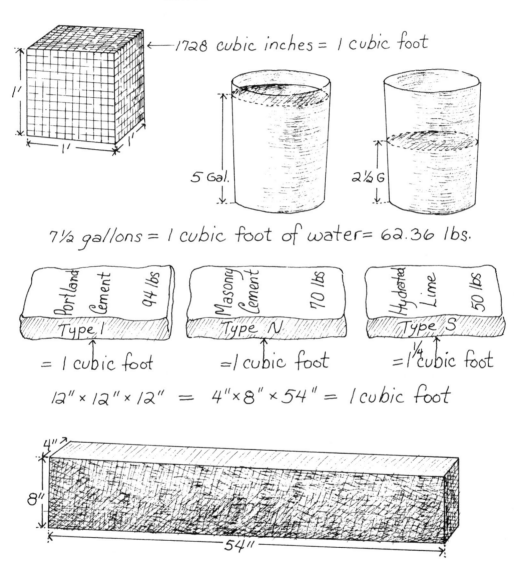

What is a cubic foot ?

1728 cubic inches = 1 cubic foot

1'

1'

1'

5 Gal.

2½ G

7½ gallons = 1 cubic foot of water = 62.36 lbs.

Portland Cement 94 lbs
Type 1
= 1 cubic foot

Masonry Cement 70 lbs
Type N
=1 cubic foot

Hydrated Lime 50 lbs
Type S
=1¼ cubic foot

12" × 12" × 12" = 4" × 8" × 54" = 1 cubic foot

4"

8"

54"

27 cubic feet = 1 cubic yard

Fig. 2-13. Different weight cubic feet.

where joints were poorly crossed. A wall of the same weight built of lower, longer stones probably would not have cracked.

FIGURE 2-16 shows a Civil War *breastwork*, a hastily built stone wall for defense purposes. If you can find stones like this, your job will be easy. More likely is to gain access to rougher stones. I often pick through old fence rows for the small percentage of well-shaped rubble similar to the stone used for this breastwork.

FIGURE 2-17 shows a dry-laid retaining wall built of rough-dressed limestone. If you can find stone like this cheap or free, you are too lucky. Stones shaped like these can be laid blindfolded and will make a very stable wall. The amount of labor represented in a wall of this sort, before it is stacked, is astounding. If you

Fig. 2-14. Broken stucco reveals nice building stones at a ruins.

gather stone and try to shape it into square shapes, you will see what I mean.

The next step in refinement is when stones are cut to fit with tiny, uniform mortar joints. FIGURE 2-18 shows a century-old stone building with several different degrees of workmanship all in the same area. Most of the stone is rough-dressed from the quarry, and impeccably fit. The cornerstones, or *quoins*, are drafted with a toothed chisel to bring the bedding planes within 1/16 of rectangular. The mortar joints between these corner joints are almost too thin to see. Yet most of the face of these quoins is left rough. The date stone is finished much finer, probably with abrasives, and has carving, which is technically out of the realm of stonemasonry. The long stones at the base of the building are finished with a bush hammer and have a uniform, pocked surface.

It is hard to believe the fuss that goes into a building of this sort. The irony is that it was built before the first pickup trucks, and now that everyone has big machines at their service, we can't have stonework like this anymore. The amazing thing is that, after 107 years, the structure is brand new, for all practical purposes.

Buildings like this will not keep stonemasons busy because they never need any work. In another century, maybe someone will get to repoint it. Buildings like this are also never demolished so you can't really have any stones like these. You're 100 years too late. More on fine stonework in Chapter 13.

The stonemasonry most available to the middle class is rubble work; that is, stonemasonry with un-cut rocks. Cutting stone to close tolerances is not a skill likely to be mastered by the weekend builder or handyman. If you could develop the required patience for squaring-up rocks, you could go a step further and do stone sculpture. The reason cutting stone for masonry has no appeal to me is because I know I can build a good, long-lasting wall with stones just the way I find them, and I can do it three times faster if I don't cut the stones. I think well-done rubble work is much more interesting to look at than ashlar masonry, although I will grant that ashlar has the edge in strength.

Come on a Stone-Gathering Trip

I needed to build a dry-laid wall, 2 feet thick, 2 feet tall, and 15 feet long. Both sides of the wall were

Fig. 2-15. It is hard to knit a wall together with chunky stones.

Fig. 2-16. This Civil War breastwork is made of good rubble.

Fig. 2-17. A rough-dressed limestone retaining wall without mortar.

to show, and the top too, so nearly all the rocks for this project would be face stones. Most stone projects of much thickness have lots of stones that are not exposed anywhere on the wall surface, but this job would require mostly face stones, or stones that have a broad surface exposed. The total cubic feet is 2 × 2 × 15 = 60. A dry-laid wall is probably one-third space, so it only weighs about 100 pounds per cubic feet, instead of 150 pounds for laid-in mortar masonry. 60 × 100 = 6,000 pounds = 3 tons. Closer to the job site, I would get about a ton of smaller fill stones and try to fill the space between the big face stones.

I decided to use mountain sandstone for the project, and I got a permit at the closest state forester's office. They ask you where you'd like to gather, and issue a permit that is good for a few weeks. The cost in 1986 in Pennsylvania was $8.00 for 2 tons, and that was the minimum size deal. Similar permits are granted for firewood gathering. There is no weigh-in or anything. It is an honor system.

I usually gather the desirable stones into little stacks at the sides of the road. Then I can back up the truck and get five or ten at a time. If you have to carry rocks through the woods, the tons will add up slowly and painfully. If you are renting or borrowing a truck to gather stone, it would be smart to have the scouting done and the heaps made before you rent the truck. Then you can come back with a truck and move the rocks quickly. A wheelbarrow would be helpful in the preliminary gathering.

Use of a Ramp

When a stone is too heavy to lift, you might find it easy to walk the stone over to a ramp, or board, then roll it up on the pickup bed. I use an 8-foot-long, 1-inch-thick oak board as a ramp (FIG. 2-19). It is amazing the abuse an oak board can take. If you're moving rocks weighing more than 200 pounds, you should use two boards and two people. When the stone gets to the top of the board, it will kick up the bottom of the board. This is no problem if you are anticipating it.

I prefer an old, beat-up truck for this kind of work. New paint and rocks are a combination that makes me nervous. Protecting the bed of the truck with a sheet of plywood isn't a bad idea, if the plywood is junk. The bed suffers the worst damage, although steady stonemasonry will destroy all aspects of a pickup truck, so don't use something too precious.

I take care not to drop the fragile stones on top of each other when I unload. The faces are easily scratched and broken. Also, it is easier to build a wall when the assortment of materials is laid out for inspection. When you are looking for a certain shape, you can scan the whole pile.

Fig. 2-18. This old stone building shows three degrees of finish stonework: rough dressed stone, drafted cornerstone, and fine finished date stone.

28

There are some very flat surfaces on some stones. This is the way they come from nature. Certain veins of surface stones are very square, and others have angles on their tops and bottoms which make them useless for stacking.

When loading a pickup truck or trailer with stone, be sure to get a lot of the weight forward of the rear axle. The tendency is to load toward the back of a truck, where it is easiest to reach, but this makes for awful steering. I usually put all the lighter rocks toward the front of the truck to balance the load.

It definitely helps to fetch rocks in pairs. Two people can lift a stone right onto a truck that would have required a ramp with one person. Because the places where stones are available is often in beautiful country, it is fun to make a family outing out of gathering them. My children are quite helpful in gathering the small fill stone in buckets. I can usually get three tons in an easy day. I call it easy because half the time is spent driving, unless the stones are very near the job site. Ten miles is the farthest I've ever driven for a tone of stones. More than that would really start to cost in time and gas.

When To Gather Stone

There are times of the year when stone hunting is best. In summer, it is hard to spot fieldstone lying on the ground because of all the foliage. Bugs and snakes are at their liveliest. In deep winter, if you have winter, stones will get frozen to the soil with greater adhesion than if they had been laid in mortar. You need a bar to pry them loose, and even then they come up with big chunks of frozen dirt. Snow will bury stones from view. If you have a wet season, that's a bad time to get off on dirt roads. Being deeply stuck in the muck with a full load of stones is worth missing. In the Northeast United States, I think the best time to gather stones is in late fall, when the foliage is down, but before there is snow and the stones are frozen to the ground, after the bugs have died and the snakes have gone to bed and before hunting season: November.

Rocks from Ruins

If you are fortunate enough to find nice stone in a wall of an old abandoned building, you will be able to build as nice a building with it. If the stone is stacked, especially without mortar, and it looks good, it is good. If the stone wasn't shaped right, the old building would be in a crumbling state. If you gather the stones from a building that fell apart, you will probably also get a building that will fall apart.

Another nice feature of getting materials from old stonework is that you can also get the right amount

Fig. 2-19. Using a ramp to move a big stone without lifting it.

of filler stones and shims. You can't really build a rubble wall without lots of fill, but it is hard to remember to gather the little stuff when you are primarily looking for big, long, boxy rocks.

Another advantage in finding a wrecked stone structure is that you can tell how much wall you can build from the amount you remove. Tearing apart a well-made wall will tell you a lot about how it was made. If the structure is 100 years or older, there will be no portland cement, unless it was added to the surface at a later date. The old lime bedding mortars don't offer much adhesion, so you can take apart old well-made walls by lifting each course of stones off the ones below. Pushing a wall over won't help the gathering game. Instead, disassemble.

Looking at an old stone ruin that is in bad shape can tell you a lot about what caused it to fail. Here are some of the likeliest reasons:

☐ Failure of the nonstone part of the building, which took some of the rocks with it—as in a fire; or a roof collapse from a snow load where the rafters pushed the walls out.
☐ Poorly built stone part—walls not thick enough; poorly shaped stones; joints not crossed; no tie stones.
☐ Bad foundation—the footing was not below frost-heave level; it was not wide enough to carry the load; or the wall was built on shifting ground.
☐ Erosion—water flowing behind, in front of, or beneath the wall undermined the foundation, causing the wall to fall into the new hole.
☐ Vegetation—tree roots got hold of a stone wall and destroyed it, vines and roots slipped in and under a wall, and got bigger and got pulled and wrecked the wall.
☐ Internal freezing—the wall was built with internal cavities that hold water (usually in dirt) and the water froze and got bigger, pushing the wall apart.

SHAPES OF STONES FOR DIFFERENT JOBS

The shape of your stones is far more important than its geological or chemical classification. Igneous, sedimentary, metamorphic—they've all been used

successfully in stonemasonry. I use them all together, which is a break with tradition.

I suppose for the best longevity, it makes sense to have all the stones wearing out at about the same time. You can tell by feel if a stone will weather well. Obviously, a big block of chalk wouldn't fare too well in a stone wall. Stones with a lot of iron tend to bleed rust down the face of walls. It will soak in and stain white mortar joints, and for that reason should be avoided. Stones of that nature will look rusty when you find them. Shale layers come apart too easily to be used as building stone. Other than these few cases, pick stone for the shape. FIGURE 2-20 shows ten different shapes of stone for ten different purposes:

1. The most forgiving type of stonemasonry is a below-grade footing for a residential building. You could gather all manner of horrible rocks and toss them in the trench, and it would work well if you made a good attempt at filling the spaces between stones with smaller ones. This is the one place you could use big, roundish stones.
2. You can use strangely shaped stones successfully in dry-laid retaining walls of a rough nature. Retaining walls can be very ragged on the side that doesn't show. Because this type of wall is often 3 feet thick, there is opportunity to use big, ugly rocks, but not round ones.
3. For a dry wall that is to have both sides show, the shape of the stone is more important and more scrutiny is needed in the gathering. No stone can be wider than the thickness of the wall; many stones should be the same as the thickness of the wall; those that aren't that wide must be considerably narrower. You can't put a very shallow stone on a dry-laid wall with any stability. If your wall is 2 feet thick and you have a stone that reaches into the wall 21 inches, the 3 inches of slack on the opposite side aren't enough to fit a decent stone. So, in stonework with two faces, certain sizes of stones are neither big enough nor small enough to be useful.
4. The next level of fussiness is for stones to be used in a mortared rubble wall. Generally, a mortared wall must fit tighter than a dry wall or the mortar joints will be too big and the wall will look pretty bad.

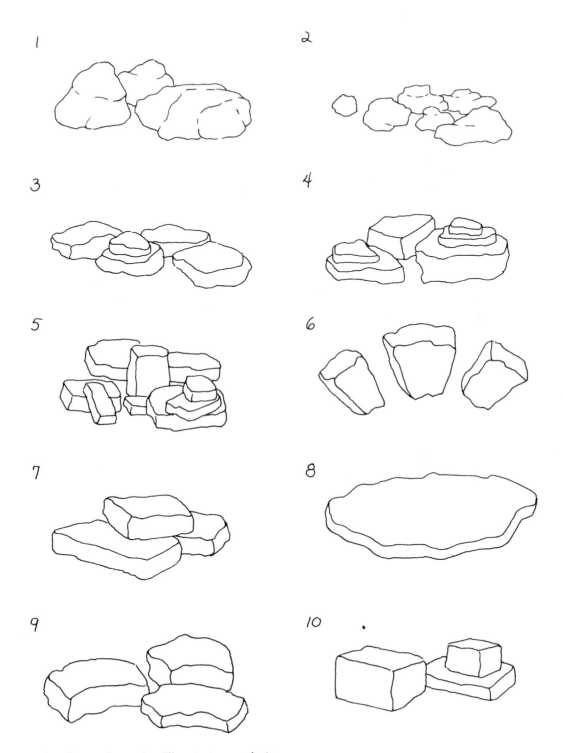

Fig. 2-20. Different shapes for different stone projects.

5. The fifth set of stones in FIG. 2-20 represent another level of refinement, which you would need for a stone-facing job. If you are facing a block wall or chimney with 6 inches of stone, then there is a new limitation on the materials: all stones must be no deeper than 6 inches. A chimney-facing job requires a high percentage of cornerstones, or stones that have exposed surfaces in two planes of the wall. Facing an interior block wall might require no real cornerstones, but the thickness limitation would remain.

6. Slightly wedge-shaped stones are handy for arches. It is quite possible to find rubble that has the right shape for a rough, but effective arch. We are used to seeing arches made of uniform units, but this is not essential.

7. Stones for steps should be big. I like a stone step that is a single rock that won't move when I jump on it. The dimensions of a good step stone might be 8 inches high, 24 inches wide, and 18 inches deep. The 18-inch tread allows a 12-inch stepping surface, and a 6-inch overlap with the next higher step. A stone of this dimension would contain 2 cubic feet, or about 300 pounds—definitely a job for two or more people.

8. *Flagstone* is very thin stone that is laid flat for floors and patios. Technically, any broad, flat stone of less than 3 inches in thickness is a flag. It doesn't matter what type stone it is. Flagstone refers to the shape. If you are going to lay a stone floor or veneer a wall, you will want all the stones to be flush with each other, so you shouldn't get any stones that are a lot thicker than the rest, or you'll be wasting mortar trying to get all the thinner stones to that level.

9. Round stone structures can make use of unsquare stone, with sides that slope inward, and even slightly outward, on the inside face of the circle. The stones should be flat unless a dome is desired, in which case the tops can slope slightly inward. An igloo of stone is a distinct possibility, but it would be hard to make with uncut stone.

10. Ashlar is at the end of this list. When you have decided to get stone into the shape you want with hammer and chisel, you've left the realm of this book. All rubble masons will knock a protrusion off a stone with a hammer or give a stone less depth by smacking some off, but this isn't the same as stone cutting. Naturally, cut stone can do all of the other jobs in the list. You could make an excellent below-grade footing with ashlar stone. You could make a superb dry-laid wall with ashlar stone. It would, however, be wasteful to use the precious stuff where it wasn't needed.

FIGURE 2-20 should make it obvious that you don't go about gathering stone the same way for all jobs. The best stone will work for any job, and the worst stone will work for none. If your project was small, it might be wise to count the number of cornerstones you need, find that many, and hold on to them until they are needed. A common tendency is to use up the best stones right away. It is hard to dole out the nice rocks on a job, but you will probably need to adopt that attitude with rubble stonework because there aren't that many great rocks lying on the ground. Of course, if you can find enough of them, there is no reason to not build a whole wall with cornerstones.

If you try the experiment of building a miniature dry wall, you'll have a much greater understanding of what shapes of stones will fit well together. The most crucial moment in which to have skill is during the gathering, and as I've said, if you can only hire a rubble mason for a day, hire him to help you gather stone. The mason will not only look at the face—the part of the rock that one envisions to be exposed to the eye—but he will picture the stone in the wall and will see the trouble it will create with the next above, behind, or to the side. You might think you've spotted the loveliest of rocks, but it could just be a headache if it won't fit in a wall.

It's alarming how delicate stone can be. We think of stones as being nearly indestructible, when in fact, most stone is something like a chunk of glass. If you think of stones as chunks of glass, you will be less surprised when you see how chipped they get from rough handling, and you might be more careful moving them. When some stone gets broken, the edge that remains on the rock is sharp as broken glass. A blunder I still commit is to grab a big piece of freshly broken stone without looking at the edge. Stone Age people made knives out of the stuff, chipping siliceous stone. Limestone rarely has a knife edge when broken, but quartz is so sharp it sometimes will cut through gloves.

3

LAYOUT AND FOOTINGS

BEFORE a wall of stone is built on the ground, it must have a *footing*, or a solid level seat that goes into the ground beyond the depth of frost. Usually, a footing is wider than the wall it supports, and spreads out the weight of the wall on the subsoil. Before you can make a footing for a wall, you need a layout to know where the footing goes. To break you in gently, I've chosen the easiest possible masonry job to illustrate layout and footings.

LAYOUT FOR A RUBBLE-STONE FOOTING

In this imaginary project, the building will be 24 × 40 feet with 18-inch-thick walls. It will be centered on a footing 24 inches wide, so there will be 3 inches of footing beyond each face of the wall. Stonemasonry starts with a pencil and paper and, hopefully, a calculator. You can use math to help find certain points on the ground, which, when connected with string, describe the place to dig the hole for the footing that will support the stone walls that will hold up the roof.

Layout on sloped land is so much harder than on level land that, for the sake of starting with the easy stuff, let's assume that your building site is roughly level. It is nice to have some extra space cleared around the site to maneuver materials, and later, to let in light. If you want 10 extra feet of space all the way around the 24- × -40-foot building, you'll need to clear a 44- × -60-foot area of trees, bushes, stickers, and rocks. If you're clearing by hand, you could do with less clearing. You can pace out this big rectangle or measure it very roughly.

Assuming you now have a clear and fairly level building site somewhat larger than the building to be,

you are ready to get some lines on the ground that will show you where to dig. The footing trench you will dig, or have dug, will let your wall bypass all the soft material that is prone to changing size and shape. A wall can't rest with stability on ground that will compress or freeze and expand.

If bedrock comes to the surface, you can build on it without a footing, or if you encounter bedrock on your way to the frost line, you can stop at that level and have solid footing. In the unlikely event that soft topsoil is 4 feet deep where you build, your footing must extend deeper to firm soil even if the frost line is not that deep. Of course, if you want to make the footings wider, you could build in softer dirt.

A footing spreads out the weight of the wall, as well as bringing it to firmer ground. If our own legs ended in spikes, our weight would cause us to push through soft dirt. Because we have feet at the ends of our legs, our weight is spread out and we don't go through the soil. To walk on soft snow, our weight requires a bigger footing: a snowshoe.

Tools for Layout

If you plan to do anything other than the smallest of masonry jobs, you should have a spool of mason's line. This braided nylon string is amazingly strong, and it stretches so it won't sag. A chalk box, the kind that doubles as a plumb bob will be helpful. A 100-foot tape and a 50-foot tape are essential. Very long spikes or sections of rebar are helpful for establishing temporary points on the ground. Wooden stakes and a big hammer are needed to establish the more permanent reference points for layout. A water level can be very useful. You also will need a hammer and a few nails to attach strings to stakes.

Know Your Diagonals

If the walls of your rubble-stone building are to be 18 inches thick, the footing could be 24 inches wide and be adequate for a one-story structure. A 3-foot-wide footing would be better, but it would also be more work and too wide for most backhoe buckets. At some point in thickness, a wall would need no additional width in the footing. A 10-foot-thick stone wall does not need a 20- or 30-foot-wide footing, especially if it is only 10 feet tall. It does need to start on firm ground beneath the frost line, however. In some old stone buildings, sometimes there is no base of stone wider than the thickness of the wall, and sometimes the footing is a few inches wider on both sides, than the stonework above.

The wall should be centered on the footing. To know where the footing goes, locate the outside corners of the building. To find the exact position of those outer corners, it helps a lot to know the diagonal dimension of the rectangle. A pocket calculator is great for this. Use the Pythagorean theorem ($A^2 + B^2 = C^2$), where A represents one side of the right triangle (or in this case, one side of your building to be), B is the other side, and C is the hypotenuse, or the diagonal. To find the diagonal of a building 24 × 40 feet, you have $24^2 + 40^2 = C^2$, with C being the diagonal. 24 × 24 = 576; 40 × 40 = 1600; 576 + 1600 = 2176; 2176 = C^2. The square root of 2176 is the diagonal of that rectangle. Punch the number and hit the square root button and presto, the answer is 46.65 feet, which is the length, rounded off to the nearest hundredth of a foot.

Turning Decimals to Inches

The figure 46.65 feet must be translated into 46 feet and a certain amount of inches, to the nearest sixteenth inch. To turn hundredths of feet into inches, multiply the hundredths by 0.12, in this case, 65 × 0.12 = 7.8 inches. Now we know the diagonal is 46 feet, 7.8 inches.

Unless your tape measure is divided into tenths of an inch, you'll have to change those tenths into sixteenths. Multiply the tenths by 1.6, in this case, 8 × 1.6 = 12.8 sixteenths of an inch, which would round off to 13/16 inch. Gathering up the fractions, the diagonal length, in terms your tape measure will understand, is 46 feet, 7 3/16 inches.

Now that you have that near sacred number, which is a lot easier to get than it sounds, you can find the corners for your job site. Firmly place a small stake at the starting point. Put a small nail in the top of the little stake to attach the end of your tape measure to. Put something visible, like your hat, next to the stake so you won't lose sight of it later. The next point on the ground will be the opposite corner, which we know is 46 feet, 7 13/16 inches from the first point. Hook a 50-foot tape on the first stake or have someone hold it tight and pull out the correct amount in the direction of the diagonal, then set the second stake. You now have two of your corners (FIG 3-1, TOP). To find a third corner (FIG. 3-1, MIDDLE), pull 24 feet on the tape from point A, and 40 feet on another tape from point B. Where those distances intersect is the third corner, point C. To find the fourth corner, pull 24 feet from point B and 40 from point A. Where those distances intersect is the fourth corner, D (FIG. 3-1, BOTTOM).

These four pins or stakes you now have in the ground represent the outside dimension of your stone walls. If you pull a string around that perimeter, your building will be a lot easier to visualize. If you aren't pleased with the orientation of the building, you can start the layout with a different approach that makes it easier to orient. The order might go like this: Put two stakes in the ground 40 feet apart in the direction you would like the front wall of the building to go (FIG. 3-2, TOP). To find a third corner, pull the diagonal measurement (46 feet, 7 13/16 inches) off point A, and pull the width distance (24 feet) off point B. Where those measurements intersect is the third corner, point C (FIG. 3-2, MIDDLE). To find the fourth corner, point D, pull the diagonal distance from point B and the width from point A. Where they cross is point D (FIG. 3-2, BOTTOM).

If the wall is 18 inches thick and the footing is 24 inches wide, there should be another rectangle 3 inches beyond the one described by the four points to show where the outside edge of the footing goes. To find the corners of the slightly larger rectangle, measure 3 inches from the other line, near the corners of each side. Entend a string between and beyond those two points. Where the extensions cross, is a corner. Follow this procedure for each corner. Put a stake in each corner of the outermost rectangle. Con-

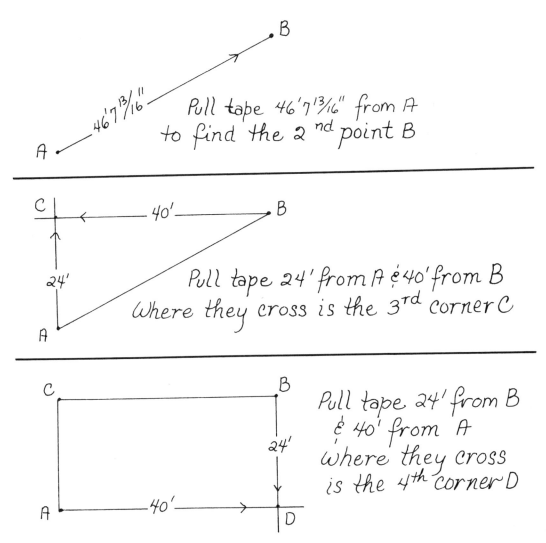

B

Pull tape 46'7¹³/₁₆" from A
to find the 2ⁿᵈ point B

A

46'7¹³/₁₆"

C ←———— 40' ————→ B

24'

A

Pull tape 24' from A ε 40' from B
Where they cross is the 3ʳᵈ corner C

C ————————————— B

24'

A ———— 40' ————→ D

Pull tape 24' from B
ε 40' from A
Where they cross
is the 4ᵗʰ corner D

Fig. 3-1. One procedure for laying out a rectangle 24 × 40 feet.

nect these new points with line to show the outside edge of the footing for your walls (FIG. 3-3).

You now have eight temporary pins in the ground, which represent the outside edge of the stone wall and the outside edge of the slightly wider footing. Obviously, all these pins are in the way if you are going to excavate for the footing. You need to establish some semipermanent reference points so that you can remove the corner pins to dig the trench, and then find those points again quickly by measuring from the reference points. The idea is to extend the original diagonal lines a set distance beyond the actual

corners and set up a plumb and solid stake at that point. Suppose 4 feet beyond the corners will get you out of the way of the trench work. You could also set stakes well within the work area a set distance from the corners, along the diagonals.

With a helper, pull a line from corner to corner and extend the line 4 feet by measuring along the line, making sure it is in alignment with the two corners. Coming straight down from the 4-foot extension (use a level or plumb bob) drive a stake in the ground. This stake won't be moved for the rest of the project, so put it in solid, with braces, if needed. If the stake is

35

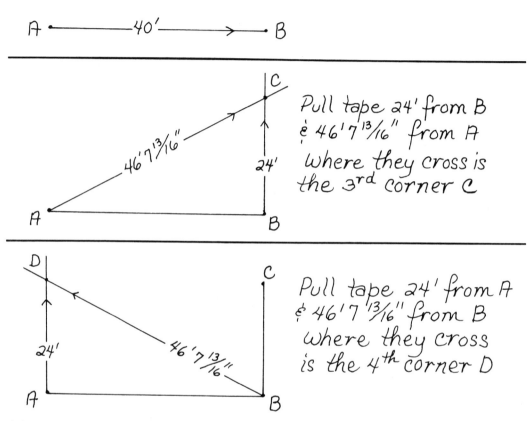

Pull tape 40' from A
to find the second point B

A •————40'————→ • B

Pull tape 24' from B
ė 46'7 13/16" from A
where they cross is
the 3rd corner C

Pull tape 24' from A
ė 46'7 13/16" from B
where they cross
is the 4th corner D

Fig. 3-2. Another approach to laying out a rectangle, starting with the long side measurement.

a piece of 2 × 4, you will need to recheck the positioning from the original corner stake, and define the exact spot on the top of the stake with a nail.

Extend the other three corners in the same way. As you set subsequent stakes, it is very helpful if their tops all come to the same level, for reasons which will become clear later. Therefore, they will all follow the lead of the first corner extension stake. The first one, then, should be set high enough that the level string you will attach to it will not touch the ground. It should not be set so high that, on lower corners of the work site, excessively tall stakes would be needed

to match the altitude. When the first one is at a good height, use a water level to set the other stakes to that same height.

A water level can be as simple as a pail of water and a length of clear tubing. The plastic tubing needs to be long enough to reach between ends of your layout, with some extra. In this case, 50 feet would be enough. Set the bucket of water next to the first stake and adjust the bucket and water level until the top of the water is at the same level as the stake top. Fill the 50 feet of hose with water and put one end in the bucket (you can gently clip the tubing onto the bucket

36

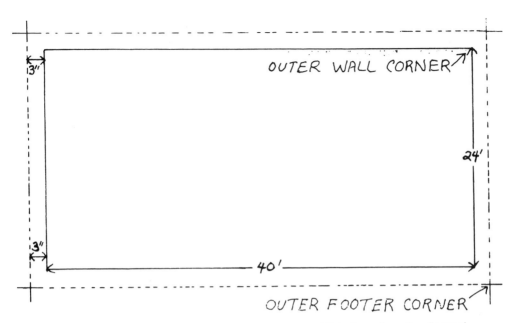

Fig. 3-3. The edge of the footing is 3 inches beyond the wall, or 4 1/4 inches along the diagonal.

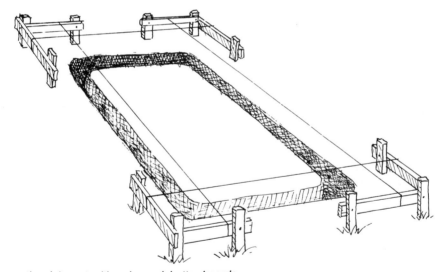

Fig. 3-4. Conventional layout with unbraced batter boards.

with vice grips or a clothespin so it won't come out of the water, but don't pinch off the tubing) and take the other end to another new corner stake, holding the end of the tube so water isn't lost. If you hold the end of the tube higher than the bucket, no water will flow out. Hold the tube up higher than the bucket and see where the column of water in the tube comes to

rest. That is the same altitude as the top of the water in the bucket. One person holds that tubing while the other adjusts the stake to come to that same point. Repeat the procedure on the other extended corner stakes.

Water levels are fun and cheap. Some tips concerning their use: Make sure there is a good flow through the tube before trusting the reading. Air bubbles will render the device ineffective. Also, if you do drain water to test the system, make sure you replace the water in the reservoir bucket to its intended level. Put a mark on the bucket at the correct water level, so you can get it right again if a dog comes along and drinks an inch. Put some food coloring or ink in the water so the column will show up better in the tubing. It is hard to see water in clear tubing. Get small-diameter tubing from a hardware store; it's cheap.

When you have these four stakes in the ground, tops all at the same level with a nail in the stake to mark the exact spot, measure from corner to corner and make sure that the positions are correct. The new diagonal measurements should be 8 feet (four plus four) longer than the original. From the extended corners, you can find every other line you will need to excavate the footing and lay the stone. You can pull up your old corner pins and find the same points again by measuring in 4 feet along the diagonal from the new reference points and marking the ground straight down from that point on the line. You will have the outside corner of the footing if you measure out square 3 inches, as before, and cross lines at the new corners.

Marking for the Excavation

Put a small stake or big nail in the ground at the point that marks the outside edge of the footing, which is 3 inches beyond the stone wall to be. Do this at all four corners, then connect the four pins with string. Next, sprinkle chalk or lime along the string to get a mark on the ground to show the place to dig. Remove the pins and string. Usually, a mark for one side or the other is adequate to make a trench. A backhoe operator using a 2-foot-wide bucket would follow the line with the edge of his bucket and the other side would be right.

If you want a wider trench or you are using a narrower shovel, you might want a mark for both sides of the excavation. If the footing is to be 24 inches wide, come in that distance squarely from the outer footing line near a corner and make a mark on the ground. Repeat at the other corners and connect the eight points on the ground by marking with chalk, lime, or anything that will show up on the ground. Extend the lines enough to get the new rectangle.

Other Methods of Layout

All of this probably sounds more complicated than it is. The math part takes only seconds on a pocket calculator, and if your site is approximately level, the initial layout is fast. The method illustrated here is actually a simplified version of a more complete batter-board arrangement, which can require up to 52 pieces of wood. The more conventional approach is shown in FIG. 3-4. These batter boards don't have braced stakes. If they needed them, you would need many more pieces of wood. A lot of buildings actually could be built with one line: the outer wall perimeter. Any other aspect you need to know can be quickly measured off that line.

There are many other possible approaches to layout. One is to calculate the diameters of an expanded rectangle and put those stakes in the ground first. You would know how far in to measure along those strings to hang a plumb bob that would mark the spot for the corner of the wall.

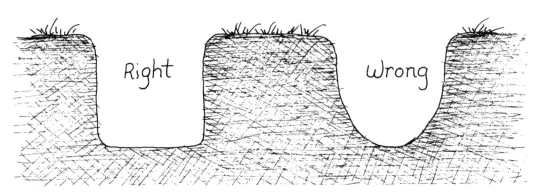

Fig. 3-5. *Keep the excavation square. Round footings can roll.*

When working with a square building, like 24 × 24 feet, you can calculate all the distances along diagonals by multiplying the desired distance you want to move the square line by the square root of 2 (which is 1.414). For example, if you had a plumb bob marking the outer wall and you wanted a mark for the outer footing, measure along the diagonals 3 × 1.414, or 4 1/4 inches. If you want the inner footing corner marked with a plumb bob, measure in from the outer footing corner along the diagonal a distance of 24 × 1.414, or 33.94 inches (33 15/16 inches).

The centers of the different rectangles that concern us in the layout of a rectangular building are all at the same point. You could use the center point with a stake to find other points using shorter tapes and walking less. If you understand the possible procedures—and there are many—the work with strings, tapes, lines and bobs will be a relief from the real work of masonry.

BUILDING A RUBBLE-STONE FOOTING

If you are young and strong, or old and broke, you can dig a footing by hand, and on a small job, you can underbid a bulldozer. The work is mindless and miserable. It is not a bad task for a work party, which is when you gather with your friends to help one of the people through a task that would be no fun otherwise. No one works as hard at these parties as they would on an actual job, but through sheer numbers, the job gets done. On many of the houses that illustrate this book work parties were used. Don't have one for anything vaguely skilled; firewood gathering and ditch digging are perfect.

The depth of your excavation will be a matter of local conditions, like the depth of frost and the type of soil. Check local codes or ask old local builders who have built things that have passed the test of time. Start your excavation at the corners, and dig straight down to the frost line, keeping the side and bottom of the excavation square (FIG. 3-5).

If you need to have the bottom of the footing level, measure down from the strings at the corners the same amount and dig to that depth. This is only possible if you had put all the diagonal lines on the same level. If your string were 3 feet above the ground and the frost line were 3 feet deep, you could measure down from the line 6 feet. For the purposes of this

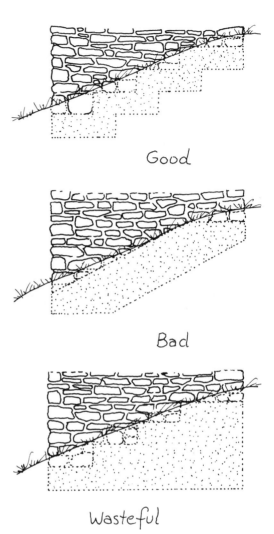

Fig. 3-6. Good, bad, and wasteful approaches to footing on a slope.

rubble-stone footing, measuring down from the ground will be adequate.

Footing on a Slope

If your job site is steeply sloped, there is no sense in trying to make a level ditch. Instead, keep the ditch level in steps that follow the grade, to save materials (FIG. 3-6). A level-bottomed ditch on a slope would waste a tremendous amount of concrete or stone. A ditch that follows the slope and remains a uniform

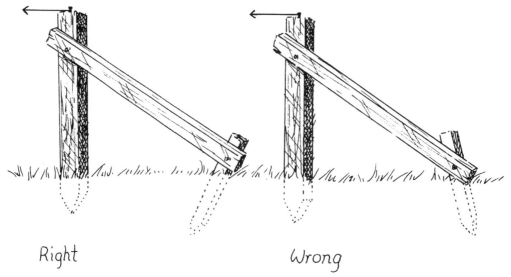

Right   Wrong

Fig. 3-7. Angle the stake to resist the dominant pull.

depth (FIG. 3-6, CENTER) would create a sloping surface which wouldn't allow stable stone stacking. The footing would tend to shed the wall in time.

In block work, the steps in a footing are done in 8-inch increments, as needed to follow the slope. The step heights accommodate the units. In rubble stonework, the step height doesn't matter; the idea is to keep a level pad on firm subsoil that steps to follow the slope of the ground so you don't have to make a giant, wedge-shaped footing.

Filling a Trench with Rocks

The very least level of skill is required to build a dry-laid, rubble-stone footing. The skill in this footing project is in the layout. The ditch can be filled with rubble stone of all shapes and sizes. The more they are laid flat in the ditch, with overlapping joints, the better the footing, but you could pretty much just toss them in and it would work. I recommend filling the ditch in layers, or courses, just to make sure there is stability. More importantly, fill as many of the spaces between stones as possible. If the ditch is filled fairly tightly with boulders and little stones, it won't shift because there is nowhere for the stones to shift to. The bottom of the footing should have the broadest stones, to spread the weight evenly over the subsoil.

For this job, you can use stones that definitely wouldn't work in a good wall. The side of the ditch

offer some support to the footing, so the integrity of the stacking is not very important. I think it is most important that the ditch be full. Frank Lloyd Wright gave a hearty endorsement to a footing for residences that was nothing more than a trench filled with gravel. A footing trench is a good place to break glass or to toss ugly chunks of old concrete. If you are someone who hates to waste anything, you can save your disposable glass containers for masonry. Broken glass also can act as gravel in concrete.

A dry-laid or gravel footing creates its own drainage system if there is an outlet pipe for the water, somewhere before the top of the ditch, that flows away downhill from the outside of the building. If the floor of the building is above the top of the footing, that drainage will keep water from appearing at the floor level.

As you bring up the below-grade, dry-laid rubble footing, try to end it at grade, or a little below grade with a modicum of integrity and using a continuous, generally level cap of concrete or mortar. The idea is to give a level, stable beginning for the wall of stone to come, which is hard to achieve using only the rough footing stones. There is no big advantage to having this footing come out perfectly level, as there would be with block or brick masonry. The first course or two of stones will get things level if the top of the footing is out a few inches. If the footing is more than 1 foot out of level from one end to the other, you should step it once. If you do any stepping of footings, remember to keep all places on the ditch bottom below the frost line, and don't create situations where the footing will be exposed to the eye. To get all of the footing out of sight on a stepped footing means you will need a little part of the stonework buried from view too.

If you are worried about the integrity of a dry stone footing, you could stop 4 inches short of the top and pour a slab of concrete to the top. If you are really nervous, you could add several pieces of horizontal rebar to the footing cap. More is always safer, in terms of building safety factors, and it would be safest for me to recommend concrete as a footing material for a stone wall, with plenty of metal reinforcing. How can I resist telling you, though, that you can also have a footing made of the junk stones you pull out of the garden for no money at all, and that it will be good enough? In the case of the 24- x -40-foot building, a concrete footing would cost about $1,000.

By timing and combining tasks, you can cheat and get a lot done. Suppose you have a little farmstead and you want a building 24 × 40 feet. One day a backhoe is doing something at your neighbor's place. You ask the guy if he has time to do a little digging at your place when he is done. This can save you the travel charge for moving his equipment. You have the footing marked out. He digs off the topsoil layer, which you instruct him to lay in rows nearby where you are planning a garden.

You also should salvage the topsoil that would be under the future building. It sure won't do any good under a building. As the backhoe is digging the trenches, you might have him heap the dirt in an appropriate spot nearby, and when the digging is done, he can quickly shape and pack the heap to create a smooth dirt hump for kids to ride their bikes over. The subsoil becomes a great toy and source of entertainment.

Anytime you contract a piece of heavy equipment, you should have all your little needs thought through before it arrives. I've had excavation work done, but forgot to ask the driver to pull a certain stump on the way out. He could have done it in 30 seconds. It ended up taking me several hours. It wasn't worth calling the guy back, because there is a high initial charge to get the equipment to the job. So if you have a tree you want pushed over or a gi-

ant rock you'd like in the yard for an artsy accent, piece together these little jobs for the driver.

When the ditch is done, if your timing is good, the loader can scoop up the stones that were gathered from the old corn field. Another method is to fill the ditches up gradually with stones from the farm. If you are downright lucky, you can get someone to pay you to dump clean fill in your footing trenches. I have paid many times for the right to dump something as harmless as old, busted-up concrete sidewalk, which would make excellent footing rubble, as would old brick or block. Don't get anything that is going to break down in your footing ditch—no wood, metal, or plastic. If you happen to see a crew wrecking an old sidewalk or roadbed, or tearing down a block building, it might be worth trying to get them to dump the stuff in or near your ditch. It could save a lot of work or money.

How Much Stone Do You Need?

To get a pretty close idea of how much rubble stone you will need to fill your trench, you need to know the volume of the ditch. In the case of the 24- x -40-foot building with a 2-foot-wide hole, if you stretched that ditch out, you'd have roughly 24 + 24 + 40 + 40 = 128 feet of ditch of whatever depth you need to dig. In my area, a 3-foot-deep excavation is needed, so the total volume of the hole is 128 × 2 × 3 = 768 cubic feet. If the depth of the trench varies, take a measurement in several places and use an average figure. Do the same if the width of the ditch is variable. When you know the approximate cubic feet of the ditch, multiple times 100 to get the approximate pounds of stone required. I'm assuming the dry fill will be one-third spaces. If you use concrete or stone laid in mortar, multiply the cubic feet by 150 instead of 100. For a dry rubble footing, the amount needed would be 768 × 100 = 76,800 ÷ 2,000 = 38.4 tons. That's a lot of stones, but remember, you don't have to be at all selective. Any stone you can pick up can go in the hole. It doesn't take long at all to gather a ton if you do not need to pick and choose.

Filling a Footing Hole with Gravel

Certainly another possibility is to fill the excavation ditch with gravel dumped from a truck. Small gravel can be directed off a concrete chute directly into the ditch and you won't need to lift a finger. The mixer truck usually costs a couple more dollars per ton of stone than a dump truck. You have to decide what your shoveling time is worth. For an extra $75.00, you could get out of shoveling the 38 tons by having the gravel chuted, like concrete. If you use a dump truck, larger stones, called *ballast* stone at some places, can be had for less money than the smaller gravel. Concrete and gravel are often dealt with in cubic yards, called simply yards. Generally,

a yard of gravel weighs about 1.3 tons, and a yard of concrete weighs 2 tons.

Bracing Stakes for Layout Lines

When using the very simple four-stake method of laying out rectangular buildings or when using the more conventional batter board arrangement (FIG. 3-4), you will often need to brace the vertical stakes to keep them from moving when the force of a tight

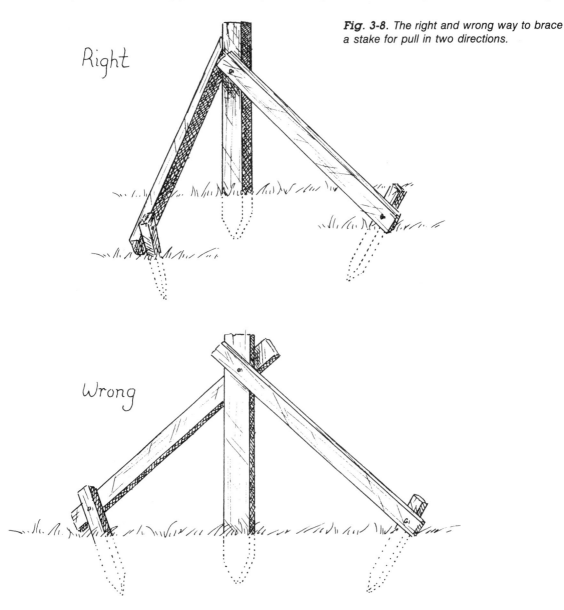

Fig. 3-8. The right and wrong way to brace a stake for pull in two directions.

string pull is applied. FIGURE 3-7 shows how you might brace the stake for one direction of pull. A diagonal piece, taken from the top of the stake to a point several feet away, is nailed to a smaller stake that is hammered into the ground at a slight angle opposing the force. FIGURE 3-7 shows how the same pull on both posts might move one and not the other. The big stakes, either used as a direct reference point or to carry horizontal batter boards, also need to be braced for a pull 90 degrees from the one shown in FIG. 3-7.

If you were to add another brace 180 degrees from the first one, as shown at the bottom of FIG. 3-8, nothing would be gained. A force in the direction of the arrow would knock over that arrangement. The correct position for a second brace is shown at the top of that illustration. This way, the post will not move from the force in either direction. The minimal layout described earlier would require four of these units for reference points.

Layout for a Straight Wall

If you were building a straight wall without corners, you would need only two braced reference points with a string between them. A chalk line on the footing or on the ground directly below that string will give you a plane to sight through, to check the positioning of the face of the stones (FIG. 3-9). Using one eye, bring the string into alignment with the chalk line on the ground, or any point along that line, and see that the stone faces remain on that plane. You can quickly check the whole wall for stones that protrude or recess too far. To establish the opposite face, measure off the first side the desired thickness of the wall.

FIGURE 3-10 shows a shortened view of a long dry-laid retaining wall with a ragged back side, which will be covered by dirt. The front side, which is exposed, was built flush and plumb by placing the stone faces along that plane described by the parallel lines of the string and the line on the ground. Materials were saved on the back side by stepping inward a little. On a wall like this, time is saved by giving no attention to the faces on the hidden side.

In FIG. 3-10, the string is also being used to determine the top of the wall. This is easy to do with a fairly low wall, and very helpful in getting a nice *cap*

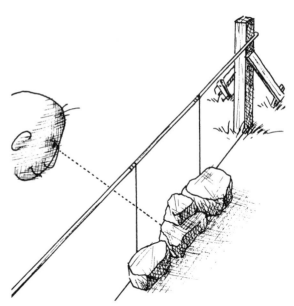

Fig. 3-9. Eyeball across plumb lines to position stones.

on the wall; that is, getting the wall to end on the same horizontal plane. Even on a wall 8 feet tall, you might want to have reference stakes that high. It would take a 10-foot 2 × 4 braced with two eight footers. The line could be moved up as you build.

In FIG. 3-4, notice how all the stakes are out beyond the work area. Any braces would be attached to be out of the way of the lines, away from the building to be. You can't hammer a stake in the ground too near the edge of your excavation without crumbling off a chunk of dirt. Everything should be back several feet or more if you expect to have a backhoe maneuvering between ditch and stakes. It is sad to have one of your references knocked over by the bulldozer, so put a bright rag on them so the operator will notice the stakes in his peripheral vision.

When you are hanging plumb bobs from horizontal lines, make sure the line is very tight and realize that the plumb bob will cause the line to sag a little. Don't let the plumb bob slide along the string. While you know it is hanging right, make a mark on the footing so you can check to see if it has moved. If the wind is blowing the bobs, you can weight the string to the right spot on the footing or ground with a brick and continue using the lines for a sighting plane to check the placement of stones. Remember, nylon

43

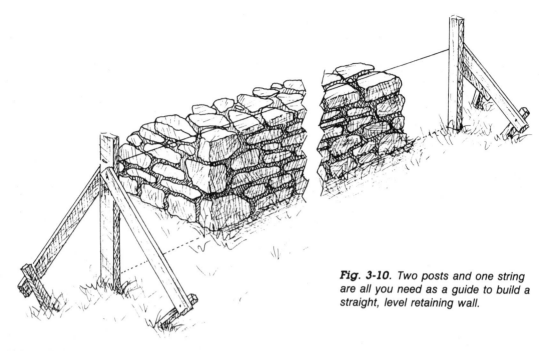

Fig. 3-10. Two posts and one string are all you need as a guide to build a straight, level retaining wall.

Fig. 3-11. A block wall with inadequate footing.

twine will stretch, so there is no sense using marks on the string itself and expecting to use them as future reference points. Strings are good for describing very straight lines that can be measured along, but the string itself cannot be a measuring device.

A low, dry-laid retaining wall that is 2 or 3 feet thick can be built without a footing. It can ride out the frost heaves. Stones might occasionally need to be bumped back in place, but that is easier than making an official footing, which would be more work than the wall itself.

For a wall of brick, block, or concrete, don't be tempted to skip the footing. These narrower walls need to have their weight better distributed or they will tend to dip into the dirt on the wet side or the low side of the wall (FIG. 3-11). Any wall built with mortar needs to have a footing or the stiff wall will burst apart.

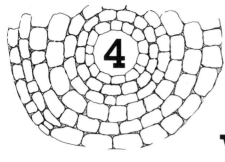

4

HOW TO LAY RUBBLE STONE WITHOUT MORTAR

TERMINOLOGY is always a problem when you are talking about a nontechnical craft with local variations. The word *point* has 37 common definitions in Webster's dictionary. In this book, I use it in two ways: as a noun referring to a tool in stone shaping, and as a verb meaning to put mortar between masonry units. It can get confusing. The thickness of a wall refers to one direction, but the thickness of a slab is a different direction. Sometimes the thickness of a wall is referred to as its depth. When talking about footings, however, depth is the direction down in the dirt. The width of a wall can also be its length. When talking about footings, width refers to its horizontal width. Square refers to 90 degrees in any plane.

FIGURE 4-1 shows a stone wall with its various parts labeled as I refer to them in this book. Imagine that there is an unseen footing under this wall that extends out beyond the wall horizontally and also goes down to firm soil.

This chapter is the heart of the book and will explain in detail how to lay rubble stone into walls. I have chosen to give this explanation without the added encumbrance of mortar. If you know how to build a stable dry wall, you will know what to do and not do with mortar when the time comes to use it. Stacking uncut stone without mortar is where masonry begins.

The very crudest sort of stonemasonry is laying a single row of stones on the ground. This seems to be a popular thing to do along roads, walks, and gardens. I've tried lining paths with stones and it did give a pleasant effect, but when the weeds started to come up between stones, I couldn't mow them or even weed-whip them because of the rocks and we ended up having overgrown paths that were a hassle to maintain until I removed the line of stones. If

your stone barrier isn't a couple of feet thick, it won't slow down the vegetation very much. A stone the size of a softball will get lost in the high grass and be a hazard. I'm not recommending you line your walkways with 2-foot-thick stone walls, I think there are better things to do with that much building stone.

THE BASICS OF LAYING STONE WALLS

Naturally, any stone wall will be more stable and last longer on a good footing. A short, thick, dry wall can be built directly on the ground with good results. The frost heaves and thaws might push stones, but if the rocks are big and well integrated, the wall will return to its original position. It can ride out the changes. With a footing, it doesn't need to.

If you need to dig a 3-foot-deep trench to reach stable ground, and you are only planning a 2-foot-tall wall, the footing is going to be most of the work. You need to dig out the dirt, move it away, and put stones in the hole. If it wasn't absolutely necessary, you could understandably be tempted to skip the footing on a project of this nature. When you're writing a book, it is easy to cover yourself by saying things like "Always use a footing twice as wide as the retaining wall is thick, and take it 1 foot deeper than the frost line" I can tell you, though, if I were building a dry-laid rubble retaining wall for myself, with 3-foot-thick walls no more than 3 feet tall, I wouldn't bother with a footing.

Footings for Drywalls

The Time-Life book, *Masonry*, suggests digging a 6-inch trench for a dry wall, and filling it with 5 inches of sand. I would be more inclined to use small gravel

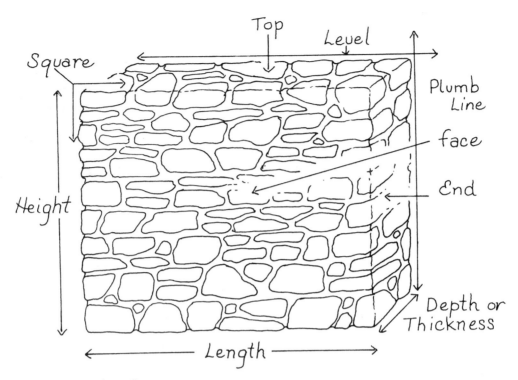

Square, Top, Level, Plumb Line, face, End, Height, Depth or Thickness, Length

Fig. 4-1. The surfaces of a wall.

for this quasi-footing, because it stays put better than sand. When I found out that ants carry off enough grains of sand to create holes in masonry, I started liking gravel better.

A compromise footing for a low, thick dry wall would be to dig down 1 foot, no wider than the bottom of the wall, and start the wall at that depth. In most areas, this would get you past the worst of the frost heave and the light, compactible topsoil.

If you plan to build a mortarless rubble wall where there is presently grass growing, you might try the following. Use a hand-operated Rototiller and till the area where the wall will be. Go over the 2- or 3-foot-wide strip of ground until the vegetation is completely tilled in with the rest of the dirt to a depth of about 1 foot. Next, remove this loose, good soil with shovel and wheelbarrow or truck, and put it somewhere that you would like a garden bed. If you mound that soil on top of an already tilled plot, you would gain the benefits of greater depth of soil. It is an easy way to get a garden started, and you don't need, or want, the soft, good earth under a hard stone wall. Inciden-

tally, while you are tilling that garden plot, pull out the stones that are kicked up and use them in the wall.

I know of no statistics, but I strongly suspect that one of the main uses of rubble masonry in the world today is to terrace steep hillsides for agriculture. Stone walls and narrow strips of level land can be found on mountainsides in the Himalayas and the Andes; continents apart. The symbiosis between stonemasonry and agriculture is also evident in the houses, walls, and barns of the old farms in the United States.

If the bottom of the short trench is filled with big, flat rocks, the wall should be okay, even if the excavation isn't below frost level. One problem with stones on frozen ground is if the sun melts the ground under the front of the wall before the dirt behind it has thawed. This will allow the base stones to tip into the soft dirt. In the Northern United States, dry walls without foundations might have trouble on south-facing slopes or in sunny openings.

Surface erosion is a greater enemy of a dry wall than is freezing ground. If water is permitted to run along the edge of a wall when it rains, in a year or

46

two, a substantial ditch will form, giving the wall a place to move into. Drainage is discussed in more detail in later chapters.

Shapes

The easiest shape of stone to lay in a wall is a domino shape, laid flat. A brick shape is good, too. The feature of being twice as long as wide is very convenient, and being broader than tall is good, too. A properly laid stone has already "fallen over" to its most stable position.

The worst shape to try to stack is a sphere. The first thing you would have to do to a sphere in order to stack it is cut off the bottom to give it a flat base. Then you would have to take a slice off the top to be able to stack anything else above. This logical progression is shown in FIG. 4-2.

Stones shaped like hockey pucks could be stacked, but at the face of the wall, there would be large spaces. If the round piece is cut in half, a rectangular surface is exposed. When turned outward, this shape could be fit tightly at the face of the wall, and the gaps in the interior could be filled with small pieces. The weakness of the wall is that the two sides are not connected with tie stones. A row of hemispheres on the top course would help a lot.

Ideal Assemblages

FIGURE 4-3 shows some of the ideal shapes and arrangements that masons lust after. With cut stone, these shapes would be realized. With rubble masonry, they are just ideals, If these stones could be obtained, walls would not have the rustic charm and beauty that natural stone shapes impart.

In an exaggeration of the ideal, for the purpose of understanding the progression of logic in rubble masonry, it could be said that you should try to build a wall of a single stone. Yet if you did, you'd be breaking one of the rules, which says to lay individual stones in their most stable position, that is, flat. This being the case, stones couldn't be higher than the wall is thick. The other limit on size is what can be lifted. Bricks, as artificial rocks, are an ideal size, because a person can lift them easily with one hand. For rubble masonry, brick-sized stones are too small because of the imperfections. They are okay for fill between building stones.

A maximum size stone in a rubble building would have the same depth as the wall, say 24 inches, a height of not more than that, and a length of more than 24 inches. The problem with a stone like this is the weight and how to move it. A chunk of granite 2 feet × 2 feet × 3 feet weighs a ton. With a ramp and

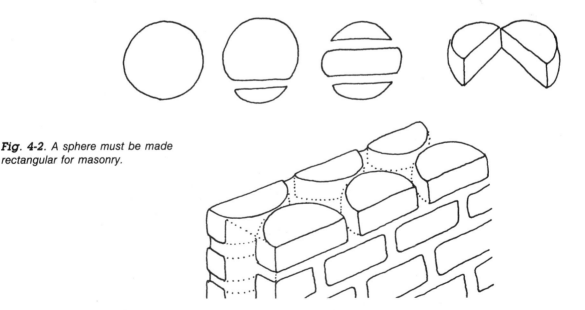

Fig. 4-2. A sphere must be made rectangular for masonry.

Fig. 4-3. Ideal shapes for stonemasonry.

Instead of large perfect stone

Can use several smaller ones

Fig. 4-4. Large square stones can be substituted with smaller units that aren't as square.

a helper, 300 pounds is an upper limit for me in stone size. A 300-pound stone would be about 2 cubic feet. This could cover a piece of wall 2 feet deep, 3 feet long, and 4 inches high. If you found a stone of this sort, it is worth extra hassle to procure it. When a big rock gets too thin, it is likely to crack in a stone wall. You might as well break it and use the two smaller pieces.

FIGURE 4-4 shows the best of stones that you are likely to find in the field. The single large stone in the corner is a bit unlikely because of its near ashlar size and shape. The arrow shows a combination of smaller and rougher shapes that can be combined to create the equivalent of the big stone.

In FIG. 4-5, the end view of the wall on the left shows an ideal arrangement of ashlar. The wall on the right of the same size is nearly as stable, although it is built of smaller and rougher stones of the sort

available to Mary and Joe Homeowner with a little scrutiny.

FIGURE 4-6 shows the next degree of roughness. To me, this kind of wall is beautiful, but it is the outer limit of funk. anything worse than this would not have much integrity. The worse the shape of the puzzle pieces, the better the puzzle doer must be. There are new types of cement blocks that are very accurate in dimension, and meant to lay without a mortar bed. Some of them have notches for interlocking. They are not much different than a set of Legos; about anyone can build with them right away. I'm all for it, but that's not what laying rubble stone is about. It is a continuous finagle to get wrongly shaped materials into a uniform whole.

I have seen walls built of round stones. I don't know why they would hold together long or how a person stacks them. To me, stonemasonry is the craft

48

of stacking flattish stones into stable walls. Spheroid-shaped stones don't make a strong wall. When they are packed in a form with concrete, the resulting wall is not a stone wall—it is a concrete wall, weakened by the addition of some oversized aggregate.

Fig. 4-5. *A wall of cut stone and a wall of field stone.*

STACKING
IMPERFECTLY SHAPED STONES

There is a certain procedure to laying rubble stone. To me, the procedure, or pattern of actions and reactions, is subconcious. I can lay a wall without thinking about the rules. To think about the rules, I have to imagine laying a wall and try to catch myself revealing the rules. It is tough to lay stone if you are thinking about how it is done. It is roughly analogous to riding a bike—hard to explain how you do it even if you know how. I need to shut my mind off and let my hands do their primitive logic. I have never seen this logic written down, except in bits and pieces. It is assembled here in the approximate order of significance.

Start at the Ends

Assuming you've made a footing or decided you didn't need one, put your stakes and line up so that you have one string that coincides with the top front edge of the low, thick, retaining wall you are building. Hang a plumb bob or set a plumb stick at either end of the wall to be and make a line along the ground that coincides with the line for the top of the wall. By sighting across the two parallel lines, you can see the plane of the wall's face. (SEE FIG. 3-9.)

Fig. 4-6. *Stones shaped rougher than these will not make a good wall.*

Starting at one end, lay a big stone on the line and make sure the corner of the stone aligns with the plumb bob that describes the end of the wall. If the rock doesn't sit quite right, it will tell you. Take some dirt away from underneath, or add a shim of stone to lift part of the rock. If the angle of the first stone, which is a cornerstone, is not quite 90 degrees, try to split the difference between the faces to keep things generally square.

Next, react to that stone with another one behind it that falls on the line on the ground which roughly defines the back of the wall. For a rough dry-laid wall, I don't think it is necessary to have a string for the back of the wall, and it can taper inward if you desire to save materials. The stone put in behind the first stone should not go higher than the first one, and the internal space between them should be filled with little stones. Make certain the little fill stones don't create a wedge effect when weight is put on them.

Because the back of this wall will have dirt against it, the face on the second stone is not important; it can be very rough. It is also fine if the back stone protrudes beyond the thickness you have chosen.

The second stone needn't come to the same height as the first. It can be shorter and the distance filled with another, thinner stone later. If you have no stones 2 inches in thickness, don't bring that second stone to within 2 inches in height of the first. You won't be able to get that part level with the first stone. So keep it one full stone's height short of the cornerstone or shorter, or bring it to the same height as the corner. Just don't get higher or only a little bit lower.

While still at one end, bring the course out a bit more, horizontally. Set a third stone adjacent with the first, no higher, and again, either the same height or a full stone shorter. Put the backup stone on the crude back line, as close to its neighbors as possible, but not taller. Fill the spaces between stones with gravel or junk stones. Don't let the fill get higher than the lead stone of the course, or it will get in the way of the flat-bottomed stones to come.

Build Up the Corner

After you have laid down a few feet of a course with both sides and the fill in place, you can go back to that same corner a start a second course of stones. The corner or end stone of that course should either stop before the one below it (horizontally), or extend beyond the one below and on to the next stone. Before you can use a longer stone to start a second course, you need to add another layer of stones to the first course to bring the height up to the level of the first cornerstone. Lay individual stones fairly level so they present a stable surface for the next layer.

After laying a few feet of one end of the first course and setting one or two stones of the next layer, go to the other end of the wall and start the other corner. Put in a big stone with a corner first, and make the next ones react to that, trying to fit as closely as possible, and not getting higher.

Fit Small Stones to Big Ones

It is easier to fit small stones to large ones than to try to fit large stones to small ones. This is one of the obvious, but subtle, pieces of the logic of stacking irregularly shaped objects. This is one of the rules, so to speak. Get the big stone in place, then react to that with smaller stuff until the area is built up to roughly the same height. Then you can put another rock on top that rests on two stones below. Fitting small stones to big leaves more options and ensures that you can fit the big stone in the wall without compromising stability by attempting to fit it well with smaller stones put in first. Let the big rocks be the bosses.

Laying rubble walls is a conservative process of laying an imperfect shape, then reacting to it with other shapes in an attempt to return things to an orderly shape. Blunders are inevitably created by the imperfect materials. The craft of the rubble mason is to constantly correct blunders with new combinations that return the wall to level, ready to accept a new stone on top. If the blunders aren't constantly adjusted and balanced, the wall will fall, right there in front of you. That's the beauty of making a dry-laid wall: it will teach you how to do it as you go, and it won't let you make many mistakes.

When looking for a stone to put in a wall, first look at the *face*; that is, the plane of the rock that will show. You will probably be looking for a fairly smooth, flat face. Check to see if the rock also has a flat base, about 90 degrees from the face surface. If so, the face will sit plumb without shims or stone cutting. If the stone has a face and a base, see how the top is. If the top slopes inward, toward the inside of the wall

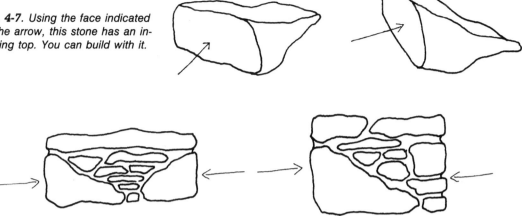

Fig. 4-7. Using the face indicated by the arrow, this stone has an insloping top. You can build with it.

Fig. 4-8. In-sloping tops seen on stones in the end view of a wall. The right side shows a maximum depth for a sloped stone.

Fig. 4-9. The end view of a wall shows an in-sloping top that goes the whole depth of the wall and becomes outsloping on the other side, and unusable.

(FIG. 4-7), it will be easy to fit in a wall. FIGURE 4-8 shows the end view of a wall that has face stones (arrow) with tops that slope into the wall on both sides of the wall. Stones are stacked between the in-sloping rocks until the internal space is filled to the height of the faces. Ideally, the stone on top of that area would span the wall (FIG. 4-8, LEFT). FIGURE 4-8, RIGHT shows a wall with a face stone that has an inward sloping top which penetrates further into the wall. this makes a trickier situation, because there is so little space left on the back side of the wall to build things up again. The problem has been solved nicely in the illustration.

FIGURE 4-9 shows another end view of a wall with a further extreme of the in-sloping stone. Here, the stone goes all the way to the back face of the wall, and creates an out-sloping surface on that face, which is not permissible. In both examples in the illustration, the stones placed above the big one will be in

danger of sliding out of the wall. The right-hand illustration in FIG. 4-8 shows the maximum safe distance that a wedge-shaped stone may penetrate a wall without setting up instabilities on the other side.

Avoid Stones with Out-Sloping Tops

FIGURE 4-10 shows a stone with an arrow pointing to the face being considered. The top of the stone would angle out of the wall if that face were used. FIGURE 4-11 shows the end view of a wall that has an outsloping stone on top. Any stone added to this stone would tend to slide out of the wall. The right side of the illustration shows a broad tie stone on top of the bad situation. A small stone has been laid on the outsloping surface and will probably fall out of the wall someday. Fortunately, the tie stone above will keep

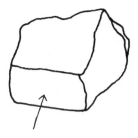

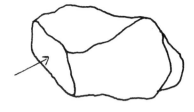

Fig. 4-10. The arrow shows the face of a stone with a trouble-some out-sloping top.

any other stones from being adversely affected by the bad shape. I'm not recommending you use an out-sloping surface, but if you do you should put a band-age over the wound.

Avoid Stones with Out-Sloping Sides

The last areas of consideration when checking a stone for its stackability are the sides. If the sides of the rock, called the *joints*, angle outward, it will be difficult to fit with its neighbors. FIGURE 4-12 shows a stone with out-sloping sides, and the same stone in a wall. The sides, which gain length toward the back of the stone, will keep adjacent stones at a distance, which will create large, hard-to-fill spaces at the face of the wall.

In rough, dry-laid work, this is rather a fine point, and the sides of the stone are the last consideration. The tighter fit a dry wall is, however, the heavier it is and the more resistance it has. A well-fit wall also looks best and houses the least amount of mice and snakes.

In-Sloping Sides Work Great

FIGURE 4-13 shows a stone with an arrow point-ing to the face. The face, base, and top are good, but the sides angle inward toward the back of the stone. This type of stone is very easy to fit into a wall be-cause the spaces it imposes are all in the interior of the wall, and can be filled. Next to the stone in FIG. 4-13 is the top view of a wall that has several stones with in-sloping sides. You can see that the vacancies are hidden from view and filled with small stone. This picture represents a single course of face stones and fill. The next course, or layer, would be staggered to connect the stones to each other.

A stone with in-sloping sides is something of a universal acceptor shape in the puzzle of rubble masonry. These stones have the advantage of accom-modating stones with outward sloping sides and main-taining a tight fit at the face. FIGURE 4-14 shows a top view of a wall containing several stones that angle in-ward. Stone A has two sides that angle in, which ena-bles it to have a tight fit with stone B, which has outward angled sides, which make it a hassle to fit. Stone B was chosen for this position because it has

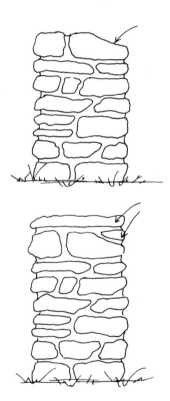

Fig. 4-11. End view shows a stone with an out-sloping top and how the damage is minimized on the right.

Fig. 4-12. The arrow shows the face of a stone with out-sloping sides. In a wall, the sides prevent a close fit at the faces.

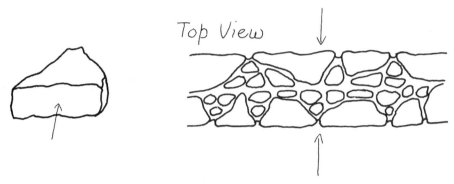

Fig. 4-13. A stone with in-sloping sides is easy to fit in a wall.

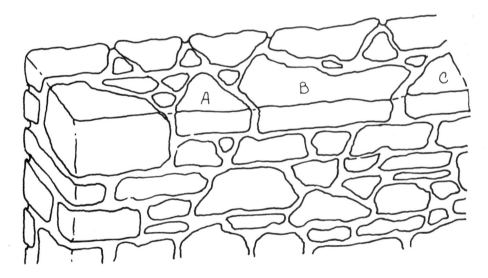

Fig. 4-14. In-sloping sides can accommodate out-sloping sides.

a long face that crosses two joints below, and has the right altitude to go with the course; in other words, its bad traits of the out-sloping sides were forgiven because of other redeemable features. Stones A and C, with their easy-to-fit, inward-angled sides, allow the blunder of stone B. If either A or C had square sides, they would have kept stone B at a distance, which would impose a big gap at the face.

Because many more stones have outward-angled sides than inward, the innies are somewhat precious. In areas of slim pickings, they might need to be reserved to adjust inevitable mistakes or problem stones.

Build Ends Higher Than the Middle

Another basic rule of the game is to start a wall at the ends, or corners, and bring the middle up to that level, not higher. the reason for this is so the easier to obtain middle stones do not impose limitations on the larger and harder to find cornerstones. Keeping the ends above the middle also allows you to string a line from end to end to have an easy guide for bringing up the middle.

FIGURE 4-15, top, shows a section of wall that has been brought up fairly level in two courses. In the center drawing, the middle has been started before the

ends. To get things well again would demand very particular cornerstones, to fit the situation imposed by the middle stones. The options have been squandered.

The bottom of the illustration shows the proper method, which enables you to use the corners with greater freedom. You are assured of fitting the important stones in the wall, and you work the smaller, easier rocks around the limitations imposed by the big end stones.

Lay Stones with the Long Side Horizontal

Another precept in the craft of rubble masonry is to lay stones flat, in their most stable posture. The logic of this is similar to starting at the ends. Laying stones flat gives you more options on followup. FIGURE 4-16, left, shows a stone standing on its end. It

Wrong

Fig. 4-15. *Start at the end of a new course, not the middle.*

Right

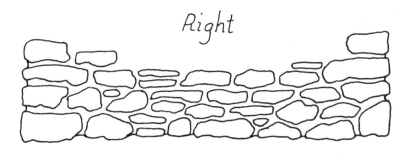

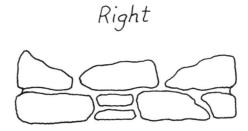

Fig. 4-16. Use stones with their long side down (right).

appears to be stable enough, but look at the very particular spaces that would need to be filled to get this wall level on top again so that joints can be crossed.

The arrangement on the right side uses the same stone to cross two other joints, rather than create joints that will need to be crossed. A stone with a short face cannot do much to cover the vertical fissures beneath it, and the continual crossing of these joints is fundamental in nearly all masonry.

Keep the Fill below the Height of the Face

Another basic rule of successful rubble stacking is to never let fill stones get above, or set limits on, the face stones. In FIG. 4-17, the fill in the wall can be seen to break the level of the course. This means that any flat-bottomed stones above this place will not sit right.

FIGURE 4-18 shows a cross section of a wall that has been violated by a fill stone getting well above the face height. The fill stone cannot be seen, yet it imposes its will, so to speak, on all the little stones that needed to be fit around it. Many more options

could have been had for the faces if this fill stone had been kept small and low. In this case, no structural harm was done. The harm is the extra time spent fitting faces around fill, instead of putting the face stones in and filling the spaces they create with small junk stone.

FINER POINTS OF STONE STACKING

There are many fine points to the rough work of rubble masonry. Some of them are rarely, if ever, dealt with in how-to books, yet it is very helpful to know about them if you want to build a good wall with the least effort.

Internal Bonding

Because in rubble work you probably won't be able to find bond stones that span the entire thick-

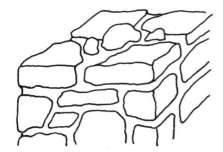

Fig. 4-17. Don't let the fill stones get higher than the faces.

Fig. 4-18. The wall is stable, but the big stone (bottom center) is a mistake.

55

ness of the wall, you will need to tie the two sides, or faces, together by overlapping the internal joint on subsequent courses. This is easiest to do when the two faces are continually brought to the same level.

FIGURE 4-19 shows a section of wall spread into its courses. Not only are the face joints crossed, the inside joint is also crossed by the way the courses overlap. Without this internal bonding, which is best achieved by laying in courses, a wall can cleave in two.

Use Risers To Connect Courses

Some coursed rubble masonry makes use of riser stones, which are stones that go beyond the height of the rest of the course they are part of, yet remain lower than the corners. The purpose of occasional risers is to make a notched connection between courses, to break up the continuous horizontal joint that is otherwise created.

If you imagine a force that is trying to shove a whole layer of wall off, it is easy to see how a riser

stone would resist that force. I don't know when this force ever comes into play, but good stonemasonry is very conservative, and between two walls otherwise the same, the one using risers will have the edge in longevity. FIGURE 4-20 illustrates the use of risers to offset the continuous horizontal line of a course, which is shown at the ends of the drawing.

Triangular Stones

Triangular stones with a plumb face and a flat base can be used well in a stone wall. FIGURE 4-21 shows a stone with a sideways-sloping top. The arrow points out the same stone in a wall. If a stone with this shape is used near the end of a wall, as shown in the top of the drawing, it becomes the problem of the out-sloping top, on the face of the end of the wall. See how little space is left to the right, to offset the wedge effect of that triangular stone.

In the lower part of FIG. 4-21, the triangular stone is farther from the end of the wall, and does not create an outward slope at the end of the wall. It takes

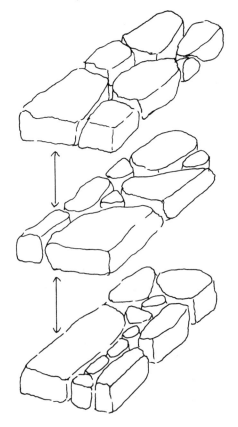

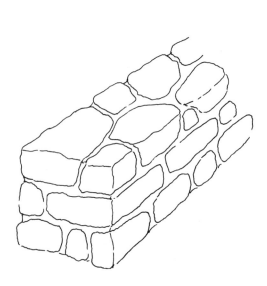

Fig. 4-19. Expanded courses show overlapping of internal joints.

56

Fig. 4-20. Riser stones break the horizontal line of the courses.

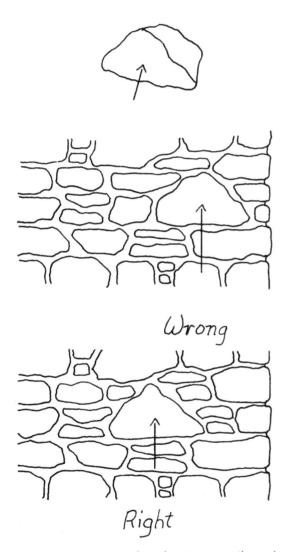

Wrong

Right

Fig. 4-21. *Don't use a triangular stone near the end of a wall where it creates an out-sloping top.*

a certain amount of space to offset the internal forces created by the angles of uncut stone.

Avoid Putting Wedges in a Wall

If you use a fill stone with a wedge shape and the bottom of the wedge is not quite touching another stone, when you put weight on top of that wedge, it will force the face stone outward. These situations can sneak up on you. Avoid using a piece of fill that doesn't quite fit between stones. It is much better to let the fill be too loose than too tight.

Avoid Narrow Stones at the Surface of the Wall

FIGURE 4-22 shows a section of wall being built. At the left end of the wall is a tall, narrow stone on edge. This is usually a bad idea, especially in a dry wall, because the stone will tend to fall out of the wall. It has little bearing surface and very little weight from above to hold it in. If a wall moves an inch from frost heave, stones with depth will ride it out, but a stone without depth can pop out of the wall.

In FIG. 4-22, there is also a rather cube-shaped stone toward the center that has gotten above the ends, which can be trouble. At the other end of the wall, there is a place where a long, unbroken joint has been allowed to develop. If you have too much invested to tear apart a goof like this, cover the weak spot with an exceptionally strong stone to keep things stable. The triangular stone toward the center will cause no problems.

If you do any fancy stuff, like make a nook or put in a round stone or a weak but interesting fossil stone, do it in an area that is structurally protected. In FIG. 4-23, the big space left under the flat stone is free to

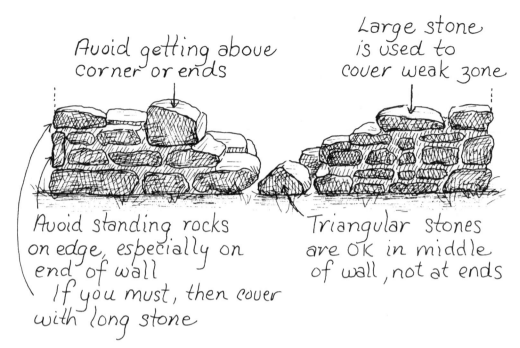

Avoid getting above corner or ends

Large stone is used to cover weak zone

Avoid standing rocks on edge, especially on end of wall
If you must, then cover with long stone

Triangular stones are OK in middle of wall, not at ends

Fig. 4-22. *A wall in progress with three common blunders.*

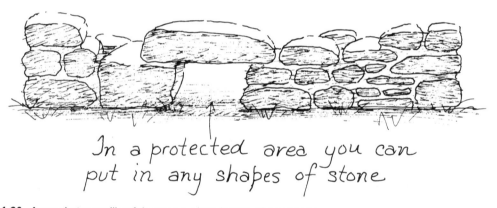

In a protected area you can put in any shapes of stone

Fig. 4-23. *A good stone will safely span a short length of weakness.*

be filled in a fanciful way, or to be left empty. It won't be a structural detriment in the wall as long as the spanning rock has sufficient tensile strength to carry the load above.

Another way to get a circle in a wall face where it will actually be beneficial to the structure is to use a salami-shaped stone that is not too big in diameter, but reaches the full depth of the wall. The transverse bonding would be worth the slight problem at the face.

Don't use a log-shaped stone near the end of the wall, because it will set up out-sloping forces on the end face.

To be really conservative about rubble masonry, you should fill even the small spaces between stable stones by laying small stones with their flat side down. The gap in the wall shown in FIG. 4-24 could be filled with the single tall stone, or more correctly, with the stack of little stones laid flat. These arrangements are like movable joint in a wall. They will absorb slight motions and spare the wall from more detrimental motion.

If you must stand stones on edge, make sure that the stone isn't composed of layers which will flake off in the upright position, that it has a very flat base, and that it is sitting well on a flat-topped stone with good depth. To be conservative, lay all your stones in their most stable position.

Rocks with Bumps

Stones that have a hump or bump on either their base or top will be a problem to fit well in a rubble wall. The bump will keep other stones from getting close and, having no real edges, the hump is nearly impossible to knock off even with chisels. On the left side of FIG. 4-25, the stone with a hump messes up the flat stone above it, by creating large, hard-to-fill spaces in the wall. On the right side of FIG. 4-25 is a more extreme situation, where a humped stone is laid on another humped stone and the bumps meet, holding the faces even farther apart. The only solution to this situation is to try to fill the spaces at the face with little flat stones, which will likely fall out of the wall, so avoid the situation altogether.

Many more stones that have been lying free on the ground for ages have humped tops than have indented tops. If it was possible to find stones with indentations, you could use humped stones with them to create very desirable combinations, but the indentations hardly exist, making the humped stones hard to fit and good to avoid. All of these shapes that I have said were worth avoiding can be worked around and used in a wall, but it is much faster to avoid troublesome shapes than to spend the time to correct the hassles they impose.

Shimming Stones

If a rock that is otherwise good in a wall has a face that won't quite sit plumb (FIG. 4-26), you can lift its base and insert a small, wedge-shaped stone to hold the base up, thereby tipping the face up to plumb.

In FIG. 4-27, stone A has been shimmed to get a plumb face, but it has brought the once level top into an out-sloping position. The stone above (C) will

Fig. 4-24. A purist would fill the space with the three flat stones.

Fig. 4-25. Stones with humps are a hassle to fit.

Shim is used to make face plumb

Fig. 4-26. An exaggerated use of the shim stone.

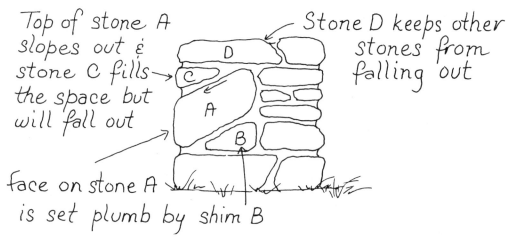

Top of stone A slopes out & stone C fills the space but will fall out

Stone D keeps other stones from falling out

face on stone A is set plumb by shim B

Fig. 4-27. A cross section of the end view reveals some of the forces at work.

probably slide out of the wall someday, but the tie stone (D) will keep any other stones above that from sliding out. Again, I'm not trying to imply that stone A is a permissible move; it isn't. What I am saying is that rubble masonry is full of little blunders and compromises. The very nature of the craft is a technological throwback. What rubble masonry has taught humanity is the importance of cut stone. so in a sense, all uncut stone is a bit of a mistake, but we can make the best of it if we pay attention to the weaknesses and keep them from spreading.

In a situation like stone A in FIG. 4-27, if I didn't have any better stones, I would split the difference between the plumb face and the flat top. Structurally, the level top is more important than the plumb face. Visually, the plumb face is more important. If you hap-

pen to be building for looks, scratch your head and go to a different part of the wall. As far as I can tell, it is impossible to do rubble masonry without having some place in the wall that bothers you and doesn't look quite right. Don't be afraid to pull down a small section of wall that just doesn't make it. It will be faster to try again for a lucky combination of stones than to keep fussing with an area that won't balance.

Rough Shaping To Make Stones Fit

Generally, only the wealthy can afford stonemasonry with cut stones. It is so much slower to shape stones with hammer and chisel than to lay a wall with imperfect rocks that you don't see much of the chisel work anymore. Often in rubble work there is a need for rough shaping, which is clearly worth

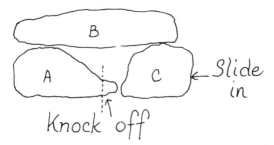

Fig. 4-28. *Rough shaping with a hammer will speed rubble work sometimes.*

the effort if it only takes a minute and allows you to fit a cubic foot of stone on the wall that otherwise wouldn't fit. FIGURE 4-28 shows a stone (A) that has a protrusion which prevents stone C from going in flush with the face of stone B, which represents the thickness of the wall in this end view. The obvious solution is to knock off the protrusion with a sharp smack of a heavy hammer, then slide C in and put the cap (B) back on. It takes almost no skill to shorten the back side of a thin stone, and that is the most frequently needed trim work. Your smack marks will not be seen, so it isn't really important how clean a break you get.

Occasionally, you might find a tie stone that is too deep by a few inches. These stones are rare and valuable in rubble work. If you are building a wall with two sides showing, it is worth going to a certain amount of fuss to shorten a tie stone so it has faces on both sides of the wall.

FIGURE 4-29, left, shows a tie stone on top of the wall that extends too far. If you were to smack the extra off in a hurry with a hammer, it might break in the wrong place and ruin your option of using the stone this way. To increase the chance of the stone breaking where you want it to, lay it in a bed of sand and score the line where you want the fracture with a ham-

mer and stone chisel. Lightly tap the chisel, and move it along the line, cutting a small groove in the stone. Do the same on the flip side of the cut. Then lay the stone on something solid, with even support under the stone, and let the part to be removed hang over the support. Give it a good solid whack with a 2-pound hammer and you should have a new stone that will fit the wall as shown on the right side of FIG. 4-29.

Bondstones

A bondstone, or tie stone, is a stone that connects two sides of a wall and allows the wall to be bonded by its weight. Bondstones are precious in rubble masonry. You will probably need to adopt an attitude of using bondstones only in an emergency—and the emergencies will arise all the time. You must use your structural kingpins where you most need them. In a sense, it would be a mistake to lay a long stone over another long stone. It is not a structural mistake, just a squandering of pieces that are in short supply and are totally necessary to hold the wall together. If your rubble stone is loaded with great bondstones, ignore this point. Chances are, however, that the rubble you can find will be shaped poorly for masonry.

FIGURE 4-30 shows the same type of stone used in two situations. The end view of a wall is shown. The wall on the right has some structural flaws at the first course. The positioning of the tie stone will keep these problems from growing. The wall on the left doesn't really need the tie stone where it is. There is nothing wrong with using it as it is used here, unless that's the only stone like it in the whole wall and there are several places where it could be used to save a weakness. You won't need to create problems for bondstones to repair; the problems will be there. They are inherent in rubble work. It's the bond stones that will be in short supply.

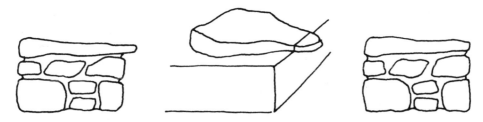

Fig. 4-29. *When your cut will be seen, score the line with a chisel before breaking the rock.*

61

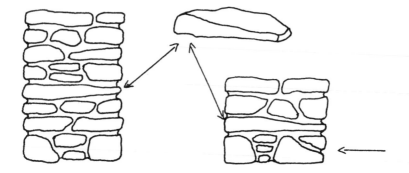

Fig. 4-30. Tie stones are helpful to bandage blunders.

FIGURE 4-31 shows a rough dry-laid retaining wall that has been in service for over a century. The 5-foot-long stone near the center spans a whole area of crummy little rocks, and gives the wall a fresh start above that point. Without that long stone, this wall would have crumbled long ago.

A stone like this is worth a bit of extra hassle for the time it ends up saving. With two people, a crowbar, a long board, a wheelbarrow, and some patience, you can obtain a 300-pound-stone. If you're doing stonemasonry, you might as well have the patience. Stones much heavier than this can really be a nuisance unless you have a bunch of people. Get some strong poles under the big rock and four people can carry a lot of weight, using the poles to get a handle and to get the body away from the rock. Some rocks simply have no handles and it doesn't matter if you

have ten people, you can't carry it without some handles. When you finally have procured a big, long stone for your wall, use it with the crappy ones in a structural strategy. If you are blessed with a lot of big, long stones, skip the strategy.

FIGURE 4-32 shows some other shapes to watch for and to use sparingly where they are needed. Stone 1 is a corner. It is hard to find stones with good right angles, so save the ones you find for places where two faces show.

Stone 2 is a capstone—that is, a stone with a broad, flat top that is exposed, as well as a face or even two on the vertical wall surface. You won't find many capstones that have faces on both sides of the wall, although masonry books always have pictures of such things to make it look easy. As you can see in the illustration, I have done the same thing. Slightly

Fig. 4-31. The long stone allows the use of many short ones below.

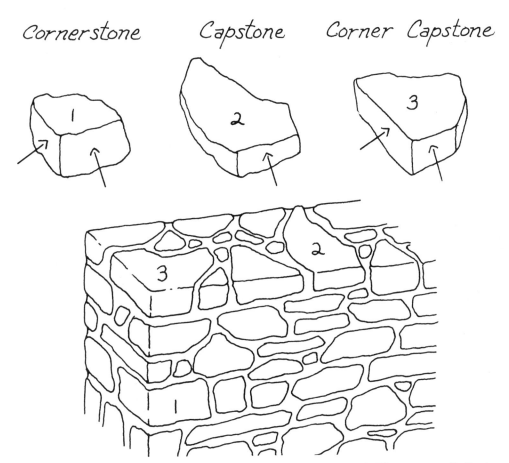

Cornerstone *Capstone* *Corner Capstone*

Fig. 4-32. *Cornerstones and capstones should be reserved for where they are needed if your stone collection is marginal.*

more realistic is that stone 2 has a very marginal face on the hidden side of the wall, and its shape was a bit of a hassle to work around. If you want the top of your rubble stone wall to be seen, you'll be using capstones worse than this.

Stone 3 is a corner capstone. This rare find has faces on three planes of the wall. Fortunately, you only need about four stones like this on an entire job. If you find a stone with a good corner and top face, save it for the top corners. The corner capstone in the drawing has an out-sloping side, to make it somewhat realistic. The small, wedge-shaped stone next to it on the right is not big enough to be very stable on the top of a wall, but there it is.

Again, your stones will probably have shapes that will force you to continually break all the rules, but you need to know the rules so that you can always

try to work the wall toward that order, in spite of the materials. For instance, it is okay to lay a stone that has a somewhat unlevel top. On the next course, you can adjust things with a stone that slopes the other way. If you don't know that it is unlevel though, you are liable to lay the next course with the same lean, until the wall is about to topple. Often, it is necessary to lay a stone that doesn't cross the joint below it. That's all right, as long as you know that a vertical weakness is developing, and you are watching for a stone to give the area a good crossing.

BUILDING A LOW RETAINING WALL

FIGURE 4-33 shows a new home and its chaotic lot on a steep hill. The gravel in the foreground is the parking space. The owner wanted a rustic stone wall to keep the dirt from eroding down the steep hill into the parking lot. The wall would also allow a more gentle slope in the yard. To accomplish this task, the wall would only need to be about 3 feet tall and 2 to 3 feet thick.

The height of a retaining wall is important in determining the needed thickness. In a sense, the rules don't even come into play if a wall is not about 2 feet tall. Up to that height, almost anything, and sometimes nothing, will hold up the bank of dirt. On very low walls, it is tempting to make them one stone in thickness. A 3-foot-tall earth bank on a hill is more serious in its demands on the wall.

Because the area where the wall was to be had just been bulldozed for the road, it was decided to dig no footing. The thick wall would begin on firm soil. The owner wanted a gently curving wall between the edge of the house foundation and the garage. To make the curves, I abandoned all layout and merely followed the whims of the situation and the rocks.

It is easy to make gentle curves with stone walls, especially if the wall is low enough that the plumbness of the faces is not too important. A wall 3 feet thick and 3 feet tall with curves can do quite a lot of leaning in and out without being in danger of toppling over. I like the curves, because then I don't have to

Fig. 4-33. The stones are at the site of the retaining wall.

Fig. 4-34. The rough stone walls transform the steep yard.

mess with lines or worry that things aren't quite straight. In this day of architecture, there are so many straight lines, that anywhere you can get away with curves, do it. Rubble retaining walls are certainly situations where you can get away with curves. Undulating walls also have more strength in resisting the ground motion than do straight walls.

Because the back side of a retaining wall is not seen, and in this case the ends of the walls are also not seen, the construction is fast. FIGURE 4-34 jumps ahead in time and shows the twin walls that terrace this hill and the set of steps that is worked into the wall. Including shovel work and rock gathering, this job took two men 5 days to complete. You get a lot of finished product for your money when you don't use mortar. The labor is much less, and there is no cement expense.

There is a trick I use on a wall like this that enables me to build the wall somewhat thinner and faster with the same strength as a thicker, slower wall. The tip is to use gravel as you lay the stones.

Use Gravel in a Retaining Wall

Gravel enables me to build a faster, heavier wall that doesn't need to be as thick as a more conventional dry wall. As I lay the big stones and fill between them with smaller stones, there comes a point when it is very time consuming to fill any smaller spaces with rocks, yet there are a lot of spaces left in the wall. If you have a gravel pile nearby, and a shovel and wheelbarrow, you can fill these spaces tightly with

gravel as you build. FIGURE 4-35 shows the beginning of the retaining wall where it connects to the concrete house foundation. The curve is the sharpest in this area, but the stones handle it well. Notice how the section of wall is highest at the end, and is stepped down so it can be tied into later.

Extra gravel is put behind each course as the wall comes up, as well as in between all the small stones. It is easier to toss a shovelful of gravel into these internal spaces than to find little stones to fill them. What normally happens is that the mason leaves the gaps empty. A wall with the spaces filled is quite a bit denser, and therefore heavier, which means it has more resistance to lateral forces, which means it is stronger, and that means the wall could be slightly thinner to do the same job. The extra gravel that is shoveled behind the wall after each course creates a wide drainage area behind the wall, thereby reducing the hydrostatic pressure on the wall from the wet earth. Removing this pressure allows an even thinner wall to do the same work of a thicker, more hollow wall without a drain system.

One of the other advantages in using gravel in a dry wall is that it is a fast way to assemble a lot of tonnage, much faster than gathering an equivalent tonnage. The walls shown in FIG. 4-34 have 12 tons of gathered stones, and 8 tons of gravel. Without the gravel, the same walls would have required another 8 tons of stone, or more. The gravel is delivered in a concrete truck, which, using the concrete chutes, can spread the small stones right where you will need them. For this project, I had them spread behind

Fig. 4-35. The front side of a curved retaining wall.

where the wall was to be built so that I could easily shovel the gravel downhill into the wall without having to throw the stone over the wall.

The first few times I tried using gravel, I had a dump truck deliver a big pile, which I wheelbarrowed into place. For a few more dollars, you can save over ten tons of nasty lifting by having the gravel brought in a chute truck. In a sense, you are hiring a truck to lay a fair percentage of the wall at a very low rate, and very fast. The gravel cost about $10 per ton delivered and spread behind the wall. For the mason to gather the extra eight tons of stone needed to achieve the same integrity would have cost the owner close to $400. The gravel backing also creates a drainage system without additional expense.

A dry-laid wall with gravel-filled spaces will probably outlast a more hollow wall. In the usual wall with hollows, the spaces have a way of getting filled over the years from soil that erodes in from the dirt behind the wall. The dirt allows vegetation to get a hold in the wall, and between the freezing and thawing of the wet dirt and the growing roots and vines, the wall can burst apart. Filling the spaces with gravel will make it hard for snakes and mice to take up residence in

Fig. 4-37. Gravel behind the wall adds drainage.

Fig. 4-36. Small stones are laid in behind the big face stones.

the wall, as they surely would if the wall was filled with spaces.

A mulch of gravel is also handy in front of the wall to keep vegetation down. A strip of black plastic will prevent the growth longer, and a cover of decorative gravel will keep the plastic intact and out of direct sun.

A certain amount of gravel will spill through a dry wall. You can minimize this action by using tight-fitting faces and by jamming dam stones in the remaining holes at the face. I have noticed no buildup of gravel at the ground in front of 6-year-old walls. I think most of the spill-through happens right away.

FIGURE 4-36 shows the back of this wall as it is being built. The big face stones are backed up with smaller rocks laid flat. Gravel was added to fill the spaces between the back stones, and make a more or less level bed for the next course of back wall stones. Never bring up the back of the wall higher than the front, because that would put limits on the choices for the next face stone. It is best when these limits fall on the little fill stones. Don't let your fill lead your faces, and don't let the faces lead the corners.

FIGURE 4-37 shows the curved retaining wall at the other end where it abuts the concrete wall of the garage. The stone does not touch the wood siding, only the concrete, which is hidden from view. From this view, it is evident that the wall is going up in crude courses, with the ends being built before the middle. The back 8 inches of the wall is gravel, which gives the equivalent of a 3-foot-thick retaining wall with much less work. Note again, that a retaining wall with drainage has less of a load to bear than one without, and the layer of gravel behind the wall will ensure that no water will be trapped in the wall to freeze and thaw and exert pressure on it.

There are many variables to consider when you are deciding how strong a wall that retains dirt should be. The nature of the soil is a big factor, as is the yearly amount of rainfall, the direction of the watershed behind the wall, the weight of objects on the dirt above the wall, and the harshness of the weather. If you plan to park a truck on the ground directly above the retaining wall, naturally the retaining wall will be dealing with more pressure than if there was only the weight of the dirt to handle.

One of the largest, and most often overlooked factors is the way in which the backfill dirt is put in place. If it is heaped against a wall by machine in a giant wedge, it will exert more force on the wall than if it is layered. Backfill put in layers in time might become self-supporting and exert no force on the wall. Natural soil is built up in layers; that is why you can usually dig a trench and have the sides support themselves, at least until the rain starts to crumble the sides. Backfilled dirt is different than naturally built-up dirt, and it is not as self-supporting when it is cut square.

If at all possible, build your retaining wall against a freshly cut wall of earth. If that dirt wall is standing on its own, obviously it is not going to exert tremendous force on the stone wall. If drainage is included, the hydrostatic forces will not develop.

Many times, the failure of a dry wall in retaining dirt is more an indication of an inadequate footing than a too-thin wall. If the base of the wall extends back into the dirt bank, it will have the weight of the dirt on top of it, and will have tremendous resistance to tipping forward. It would need to push that whole column of dirt out of the way before it could tip forward. Many walls fail because erosion cuts a hole in front of them that gives the bottom stones a place to fall to. Failures of this sort are no indication that a wall isn't strong enough; they indicate a drainage failure.

Walls that curve into the bank of dirt they retain are considerably stronger than straight walls. The obvious exaggeration of this would be a circular wall. At any point outside the round wall, you need to push the whole wall to get a stone to move. FIGURE 4-38 shows a pit I dug with a pick and shovel in less than a week one winter. In two easy days, I was able to retain the hole with courses of dry-laid stone. The circular wall is only one stone thick—less than 1 foot in places—yet it was stable enough to climb up and down at 12 feet high. I parged the inside of the pit walls with a lime mortar mix, and got a cistern that gathers rain water from the roof and holds more than 3,000 gallons. The cost was less than $20. Digging the hole was more fun than I had imagined. FIGURE 4-39 shows the pit stoned in.

A square hole of similar capacity would have required 2-foot-thick walls of stone to be sound, because there is no arch effect in the forces behind a straight wall. I dug down to bedrock for this cistern (I had planned to go deeper), so the walls of stone have a good footing. I laid the walls dry to get the hole locked in place before it had a chance to collapse. As rough and sloppy as this structure is, because of its position and shape, it will probably outlast all the other stonemasonry I've done.

Fine Points about Retaining Walls

Often the word *batter* appears in books about masonry. A batter is a slope or taper in a wall, usually from thick at the bottom to thinner at the top. There are different recommended batters for different purposes. Some of them have the purpose of saving materials without having a narrow base. On a low wall, there isn't much reason to use a batter. Chances are the courses will be too few and the stones so clunky that there isn't much chance of making a convincing taper anyway. For layout and building ease, keep the face plumb. The back of the wall could taper inward slightly to save some materials. Battering a wall makes more sense on a large concrete wall, where by merely changing the angles of the forms, you can save thousands of dollars.

Some people lean a wall against a dirt bank. I could find no statistics about the longevity of different walls with different batters, but in the search for information, I found evidence of batters that tapered

Fig. 4-38. The walls of the hole need to be retained fast.

Fig. 4-39. The cistern is stoned up quickly with one layer of dry stone.

toward the dirt, away from the dirt, and in different angles than the front of the wall. My conclusion is to build plumb walls with drainage, and to use big flat stones on the bottom course on the ground which should at least have the topsoil removed.

FIGURE 4-40 shows a section of an 8-foot-high, century-old dry-laid retaining wall, built of some pretty bad stones. You can do this too. Can you think of anything else besides a stone wall like this, that can be built entirely with discarded, in-the-way materials with no tools at all, and that will give lasting and beautiful service? This is ancient trash being put to use. These stones probably came from the cleared area around the house, so the wall helped make the lawn.

Dry stone walls are the epitome of the folk craft of rubble masonry. Many old villages depend on retaining walls to hold back mountain dirt and create level places to live. Some towns depend on dry stone walls to hold back the sea. The average homeowner might want a retaining wall to hold an earth berm against a wall of an earth-sheltered structure or to terrace some garden space. Sometimes it is worth building a dry wall just to have a place to put excess stones.

Dry rubble stonework is the one type of masonry that is likely to bring success to the beginner. Having none of the confusion of mortar can give you confidence when stacking stones. It is the mortar and the trowel that give a professional mason his legitimate look, and that's the part that will make a novice feel clumsy. Crude, dry stone walls put us all on the same level. You can build something with only your hands and the rocks at your feet, and it can be as interesting as a giant, clumsy puzzle. If you solve the puzzle, you get to keep it that way—for 100 years if you solved it right.

FIGURE 4-41 shows two of many types of batter for a retaining wall. The wall on the left leans into the bank. One of its problems is that it leaves ledges for the frost heave to push against and overturn. The wall on the right is tapered away from the dirt bank, though it has a plumb face. A frost heave will tend to ride the dirt up this wall without damaging the wall.

Weep Holes

Weep holes are spaces, usually near the bottom of a wall, that extend through the thickness of the wall to allow water free passage through the barrier. In a dry-laid wall, weep holes are not needed because the wall is full of holes and will not hold water. If you ever mortar the wall, which would be a waste of time if the wall has no footing, you would need to provide weep holes. You can make them by using in the wall an occasional section of pipe or wooden dowel that has been greased so it can be pulled out when the mortar has set.

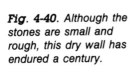

Fig. 4-40. Although the stones are small and rough, this dry wall has endured a century.

Fig. 4-41. There are many different approaches to batter.

Fig. 4-42. Extra deep stones add the weight of the backfill to the wall's stability.

Thickness of Retaining Walls

Stones that reach into the retained earth give a wall an extra measure of resistance to being pushed over. Retaining grillages along steep highway banks are successful with relatively little mass because they reach far into the bank and use the weight of the dirt to make them stable. If you leave big stones protruding out of the rear of the wall into the earth it is retaining, the weight of the earth on top of these stones give the wall the equivalent of much more mass. If you use this trick, make sure the protruding stones

are strong and not apt to crack, and that they extend through most of the wall's thickness. Also use the protruders only below frost line, as shown in FIG. 4-42. For this wall to be tipped over, either the protruding stones would need to be snapped off, or the dirt would need to be pushed out of the way, all the way to the surface.

A retaining wall should be one-third as thick as it is tall. A 9-foot-tall wall should be 3 feet thick. This rule does not apply to low walls. A 3-foot-tall wall must be more than 1 foot thick. There is a minimum thickness needed to allow a balanced structure with rough stones, probably 24 inches. With very bad stones, you might need 36 inches to do a comfortable wall. That thickness would be okay for up to 9 feet in height. A 12-foot-high retaining wall would need to be 4 feet thick, and so on. This implies that the pressure from the dirt gets greater with more height. Actually, it doesn't. The pressure at a certain depth causes the soil to act more like a solid than a liquid, and it can be self-supporting and not exert any force on a wall at great depth. Even if you were holding back sand, it would not behave as a liquid. If you were retaining water, the pressure would increase with height.

A good retaining wall should have a footing half as wide as the height of the wall. The taller the wall, the more you need the footing. A good 10-foot-high dry rubble wall should have a footing 5 feet wide. The wall itself should be slightly over 3 feet thick. With good stones, you could build a 2-foot-thick wall as high as 6 feet, but the footing would need to be 3 feet wide (FIG. 4-43).

Sloped Walls

A *slope wall* is one made up of stones laid along a steep bank to keep it from eroding. These hardly

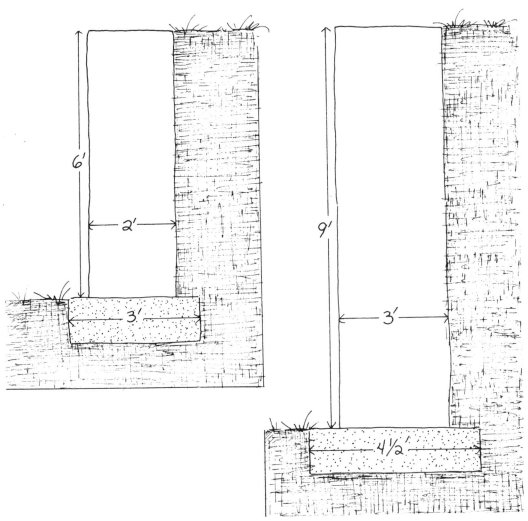

Fig. 4-43. *Ratios between height of wall, thickness of wall, and the width of the footing.*

qualify as stone walls, yet they must be started at the bottom, with higher stones leaning on the lower ones. Mountainsides are frequently held together this way naturally. Don't remove stones from steep slopes if they are all that is holding the soil in place.

Capstones

Capstones are very important in dry-laid retaining walls. Small stones are prone to being knocked off the top of a wall, and this can be dangerous if people are on the wall or below on the ground. A wall capped with heavy stones can handle children running along its top. Look again at FIG. 4-34. See how the wall stops with fairly large stones that come roughly to the same plane. This takes a bit of planning, which on a job like this is nothing more than walking to and from the rock pile and the wall with a measuring tape and seeing how much space you have and what size rock there is to fit there.

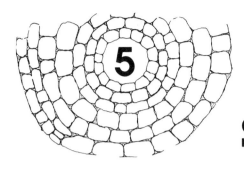

5

SMALL STONE DAMS

BUILDING a stone wall in a creek to create a pool and falls is about the most fun you can have with stonemasonry. Kids build dams every summer in the stream that goes through our town. Nobody ever showed them how to do it. They don't have any tools at all. They don't spend a dime. Yet they are able to construct a legitimate 3-foot-high dry-laid stone wall with stones from the creek, and they get a swimming pool for the summer.

Even tiny trickles can be dammed for wading pools. The stream in FIG. 5-1 is only a few inches deep and a few feet wide. It hardly seems worth fooling with. Two people decided to put one hour into building a tiny dam and now there is a pool deep enough to swim in. Ask a small child if 18 inches of water is enough to play in.

The dam shown in FIG. 5-2 is so low that it needed very little thickness or integrity. It was literally thrown together from stones that were lying in the pool area. A single piece of plastic, 15 feet × 2 feet was stretched across the upstream side of the stone barricade, and then gravel and small stones were raked against the bottom leading edge of the plastic to keep it down. The small stones also stop leaks where the plastic doesn't mesh with the creek bottom. Raking them into the dam left a more comfortable floor in the pool.

FIGURE 5-3 shows the same creek as in FIG. 5-1, with the help of the little dam from FIG. 5-2. Even though the pool is not deep, rubber raft can be ridden quite a distance into the shallow water. Close to the dam, there is enough depth and width to cool off and paddle around.

THINK BEFORE YOU BUILD A DAM

Before you do anything to a stream, even if it is on your own property, it is important to envision where the water will go when you obstruct it. Will it spill out into a low area and create a stagnant pond? Will any valuable foliage be drowned by the rising water? Will the dam create a worthwhile impoundment?

A 3-foot-high stone wall and sheet of plastic can bring about some major changes in the immediate environment; therefore, a thorough investigation of the ramifications is in order. Every situation is different. I would reject some dam sites because the water is too toxic to purposely invite swimming. Other settings are too pristine to alter. There are many situations where a temporary dam would be illegal. The stream I live by is stocked with trout, so my dam, or temporary impoundments, must allow free passage for the fish. The dam can't be higher than 3 feet, and the falls must drop into a pool of water, so the fish have somewhere to land and so the falls aerates the water.

It is good to know your local codes before you alter the environment at all, and especially a waterway. The way we are all connected is rarely clearer than along a river. What you do to the water will affect things downstream, and people upstream will affect your water.

Proper Dam Height

Generally, if there is a good flow and the stream is rocky, a dam under 36 inches will not drastically alter the situation. This sort of dam probably won't exceed the changes normally encountered in high wa-

Fig. 5-1. A shallow stream can benefit from a small dam of stones.

ter, or when log jams dam the creek. I wouldn't recommend a dam higher than 3 feet for several reasons. One, much greater pressure will be encountered from more altitude of water, requiring much thicker walls. Two, the water falling from a greater height will erode the downstream side of the dam quickly. Third, the fish will have trouble negotiating more height than that.

If your stream is prone to flooding, there's not much reason to think of permanent dams. In my situation, I rebuild the dam at the beginning of swimming season after spring flooding, and then take it apart in the fall so leaves can wash through. Once the rocks are gathered, the construction is fast. Temporary dams will cause much less trouble with the authorities than concrete dams, and will cost much less money.

There's no point trying to make a permanent dam where there are flash floods. You might as well just throw money in the steam. A temporary, dry-laid dam is made to blow apart during a torrent. When things settle back down, you can reassemble the wall quickly.

Know the watershed of the stream you intend to dam. What kind of industry or agriculture takes place above your place on the watershed? How will your dam affect the water quality downstream of your pool? Will your slowing of the flow cause the stream to warm too much? Will it prevent fish migration? Will it drown trees? Chances are, with less than 36 inches of rise, nothing detrimental will take place. The impoundment won't put the water level above frequent flood levels, and the waterfall will aerate the water. As long as the water remains within its banks, there shouldn't be any significant warming, and if the banks are relatively steep, there won't be any broad, shallow, swampy areas created by the dam.

Environmental Considerations

It is more likely that you will have a tiny creek or a small spring branch on your property than a full-fledged waterway. It is amazing how quickly a trickle can become a charming little pool with its miniature waterfalls. The stream shown in FIG. 5-1 is narrow with steep banks. FIGURE 5-3 shows the same stream after a small dam was built. We gained much more depth of water without increasing the width much at all. It is important to keep the surface area of the waterway about the same in order to minimize evaporative loss and to avoid creating a shallow, broad, mosquito-breeding pool. Look carefully at the shape of the channel of your waterway to see if you will be able to increase depth and volume without increasing the surface area a lot. If you can, the fish will be happy, and you will minimize the negative environmental impact of your project.

The height of the dam will be limited by the height of the banks. Suppose you wanted to create a pool of water in your rain gutter, which is somewhat analagous to a creek. If you build an obstruction of greater altitude than the sides of the gutter, water will spill out over the sides. If you lower the obstruction the correct amount, the pool will fill to the top of the sides, but flow over the obstruction and out on its normal course.

Suppose that same gutter varied in width from 1 inch to 1 foot. If you could put the obstruction at the narrow place, very little material would be required. So look for a narrow place in the channel. Suppose, as well, that the sides of the gutter varied a lot in altitude. The height of the pool (or depth) will be limited by the lowest spot along the sides, where the water will spill out. In the creek or the gutter, you won't want to create a new channel for the water. Stay within the natural banks to be safe.

ONE-HOUR DAM

Assuming you have spoken to the various authorities, and thought through the ramifications of your small project, you can now decide where the dam will go. You can use a level and a helper to get a fair idea of how high a dam you can make.

Determining the Height of the Dam

Suppose you'd like to raise the water level 2 feet at the dam site. Measure off the water surface 2 feet, and hold the top edge of a 4-foot level at that height. It helps to have a stick or other prop to steady the level at that height.

Then have your helper go upstream and hold a stick as plumb as possible by eye, in the water, while you sight along the top edge of the level to that stick, guiding your helper by voice to move his finger up

Fig. 5-2. A one-hour dam creates a fun place to cool off.

Fig. 5-3. The same creek as in Fig. 5-1, after the little dam is in.

or down on the stick until it is at the same level as the proposed dam. If you build the dam that high, it will back up water to that height on the stick. If that point on the stick is obviously above the height of the stream banks at that point, the water will spill out. The dam cannot be that tall without rerouting the water. If the point is below the height of the sides, that dam height will work, provided the sides don't plunge lower closer to the dam. To determine how far up the creek the water will be impounded by the proposed dam, have your helper keep walking upstream while you sight along the level until the water level itself is at that height. The water will back up until it becomes level with the dam, and then it will spill over the dam.

For a very low dam (FIG. 5-2), almost no integrity of masonry is required because there will be very little pressure trying to push the dam over. On a narrow stream such as the one in this project, you can built a low dam in a flash, provided there are rocks in the creek, and there usually are.

We merely piled the rocks up, with no strings or levels. Having water where you work is great for determining level. You can measure off the water's surface with a stick to keep the wall coming up evenly. There's no reason to build more than you need, and that's the only reason to pay attention to level on a project like this. We had the stones crudely assembled in 1/2 hour, spent a few minutes filling some of the big gaps with small stones and then stretched a scrap of 4-mil polyethylene across the wall. You can immediately feel the water pushing the plastic against the stone obstruction. If your stones aren't sharp and the gaps aren't too large, the plastic will work fine for a dam lower than two feet. With more height, greater pressure, sharper stones, and larger holes, the plastic would get torn and pushed through the dam, and the water would leak under the dam instead of increasing in depth.

Slow the Stream Gradually

It is important not to completely shut off the flow of water downstream while your pool is filling. When the plastic is in place, look downstream. Is the flow dramatically decreased? Are crawdads suddenly stranded? It is important to provide a controlled leak in the dam while it is filling, to avoid killing fish below the dam. Keep the plastic pulled up in one spot to al-

low some of the flow to continue. When the water starts to go over the dam, you can patch this hole in the bottom by pulling the plastic back down over the leak. At this point, the original volume of water should be plunging over the little falls, putting tiny air bubbles in the water.

If there is a dramatic reduction in the overall flow of the dammed creek, even after the pool has filled, then water is probably leaking into the banks, which had been high and dry and could soak up a lot of water. If the flow doesn't return to the way it was in a day or two, there might be a hole that much of the water is sinking into. You can find the hole by feel and patch it, or you can dismantle the dam.

After you put the plastic across the stones, if you haven't noticed an obvious rise in water level within a few hours, your dam is probably leaking too much. Use stones to close the little holes along the bottom of the plastic. You can feel where the water is leaking. If you muddy the water in the pool, you can see where the muddy water comes through the dam and close the hole. If your dam leaks as much as the total flow of the creek, your pool will never fill. Worse is that the dam will act as a filter, catching all the crud that comes down the creek and depositing it at your pool site. If there is a falls, the crud will spill over the top, leaving the top of the water in the pool clean.

Once the water is dammed, look around to see if there are any negative aspects to the new pool. It is a simple matter to let the water back out if there is something wrong. In fact, you should let the water out occasionally and allow the silt to pass through. Otherwise, it will tend to accumulate in the pool and cause the water to get cloudy when it is stirred. Because my pool is mainly for swimming, I break it down at the end of that season. You might want a pool for skating and take it down in the summer. You might want a pool to raise fish, in which case harvest and cleaning the pool might coincide. Some people like a small body of water just to be there. Whatever your reason, the pool bottom will probably accumulate silt and will need an occasional flushing.

If you have trouble picturing how to obstruct a flow of water, you can make a model of the situation in a sandbox, with a strip of plastic. You can re-create the circumstances of your stream bed to see how the water goes. Use plastic wrap to line the sand.

TWO-DAY DAM

Damming a larger stream or making a taller dam complicates matters considerably, but the task is still within the grasp of the weekend handyman. The dam shown in FIG. 5-4 takes two days, rather than two hours, to construct. If the materials were all there, it could be done in one long day. About once a year, we get a flood that turns the stream into a torrent, and it spreads the dam down the creek about 30 feet. It takes a few hours to gather it back. The swimming pool in the creek takes less maintenance than a regular swimming pool, so I keep putting it back together.

There isn't much you can do about stopping the flood unless you want to get the Army Corps of Engineers in on the project. So if you have a waterway that rises hard occasionally, it is best to think of your dam as temporary. There really is no way you could build a dam for those flood conditions, so you only need to make something that can handle normal conditions, and assume that part of it will topple when the wave hits it, which is all right, because you can put the thing back together again pretty fast.

A 3-foot-high dry-laid wall needs to be built with integrity. In holding back water, the pressure is much greater with a 3-foot depth than a 2-foot depth. You should observe certain rules of building stone walls, although you do not need to pay attention to the looks of the thing.

Building a dam level and straight is only useful in that you can minimize the materials. A straight wall covers the shortest distance and uses the least materials. A level top will follow the level pool of water, and you can make a falls at the desired place by pulling a stone or two. An unlevel top is fine too, except the parts that stick up were unnecessary.

One line would be very useful: the line that crosses the creek at the top of the wall to be. Place stakes into the banks at either side, and pull a string level between them. To check the level, measure off the water if there is any where you are working. If not, use a level to get the string right. (SEE FIGS. 3-9 AND 3-10.)

If your stream is gentle and has a low volume, the force of the flow will not be much of a factor, and you can build as if the water was not there. When you are working with a stronger flow, it helps to have a culvert of some sort to handle the flow while you build the wall. It should be at the lowest place in the creek bed.

FIGURE 5-5 shows the beginning of a dry-laid dam. I had some old cement blocks, so I used them as face stones to speed the thing along. The culvert is a 55-gallon drum with the ends cut out. It is a convenient 3 feet in depth, the same as the base thickness of this dam. If you look closely, you can see the single string that marks the top front edge of the dam. It doesn't look plumb with the face of the dam because of the camera angle, but it is. Notice how the first course of blocks is fairly level with the water. It is nice having such a reliable reference when setting the stones.

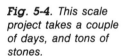

Fig. 5-4. This scale project takes a couple of days, and tons of stones.

The upstream side of the dam is the one that needs to be tight fitting and flush. The back side can be very ragged, as long as stones don't get above the face. The base should be the broadest part of the wall. To get this wall going, set one block with a level, coming down plumb from the string. Subsequent blocks or stones can be set by eyeballing across the line until it aligns with the front edge of the tested block. All other faces are set to that plane.

For the dam in FIG. 5-5, I set the culvert like a stone. I allowed water to run through the pipe so the depth didn't change while I stacked the wall, which was convenient, because it is a hassle laying stones in 3 feet of water. In this case, the water would have been deeper than that without the culvert, because the dam was begun in 18 inches of water.

The pressure will be greatest at the bottom of your dam. The bottom should be the tightest fitting and the thickest part, especially in the center of the creek bed, where there is already some depth of water. As the wall gets higher, you can make it thinner, with bigger holes. To kill to birds with one stone, gather rocks from the future pool bottom to use as fill in the dam.

If your rocks don't fit very well and you don't have old blocks, you can still build a dam. you can use discarded metal roofing to cover the face of the rough wall and prevent water pressure from forcing the plastic through the big spaces between stones. In fact, the metal roofing might work well enough to make the plastic unnecessary. It all depends on the volume of the flow.

Obviously, if the entire flow of the stream is no more than what a garden hose would put out, your dam can allow no leaks if you want that volume of water to spill over the top of the pool. If the leaks under and through the dam are equal to the entire flow of the creek, the pool will never fill and the obstruction will act as a screen in the stream, catching all the debris that can't squeeze through your dam and depositing it in the would-be pool bottom.

The bottom of the creek in FIG. 5-5 is rounded. The first job was to level out that roundness with stones. I built the area up using an 8-inch stick as a guide. Everything was kept 8 inches below the water surface, so the first blocks would come to water level.

Using the water itself as a reference will not work if your dam is stopping the flow and making the wa-

ter steadily deeper. Don't worry about exactness. You could do the entire project without a string or level or measuring stick, but the devices will save lifting stones you don't need.

This kind of work should be fast and fun. It is one of the few jobs kids will volunteer for. We use inner tubes to support plywood, and the kids can load stones on the raft, and float 500 pounds of them in 6 inches of water. Stones are lighter underwater, and it is surprising what large stones can be moved, if they don't have to break the surface. I have used a wheelbarrow underwater successfully to fetch stones that I never could have moved on the ground.

The spaces between the round culvert and the square blocks were filled with little stones. When you make the base across the creek, you can bring up the face quickly, and quite a bit above the back side. Then you can go behind the dam and fill the back. You can do the back very quickly if you have done the face with care. You can almost toss rocks against the dam in a heap. Less of them will be needed if you place them carefully, and use their bulk in the most effective fashion.

FIGURE 5-6 shows the back side of the dam before it is filled in. This couldn't be done without the blocks; the wall would have to be built front and back, course by course. The blocks wouldn't work this way without the broad, level footing. They would lean into the soft bottom sand and tip over. The method I used here also wouldn't work without the culvert. The pool would start to fill as you were building. The pressure would be building as well, and it would blow apart a wall of dry-laid block without the backup mass. If you don't have junk block or a culvert, you will have to build the wall with greater care.

Notice that the creek in FIG. 5-6 is moving through the dam and is still at the height it was when I started. If I were to block off the culvert with a lid, the creek would start to back up. Notice that there are only four courses of 8-inch blocks, plus a few inches of masonry above the water level. This gives a maximum of a 3-foot rise, which is about all this type of construction can handle.

I have done some experiments with dry-laid block and water pressure. A water tank can be made of dry-laid block on a firm, flat surface with nothing but a layer of 4-mil polyethylene plastic for waterproofing. A tank built of 8- x -8- x -16-inch blocks, four blocks

78

Fig. 5-5. The base of the dam is built around the culvert.

Fig. 5-6. The back side of the dam is built fast and rough.

Fig. 5-7. The flat face has enough backing now to handle the water pressure. The dam is ready to fill.

tall, three blocks wide, and five blocks long will hold almost 500 gallons, and can be covered with a single sheet of plywood. A water tank five blocks high without mortar will heave apart from the pressure. It is amazing that at 3 feet in height, water can be contained by almost anything, including a plastic garbage bag. At 4 feet of height, suddenly you need reinforced concrete to contain the water. Take a look at the big kiddie pools for sale. It takes only the flimsiest of framework to contain 2 feet in altitude. The diameter of the pool is irrelevant as far as the strength required. You won't see any flimsy tanks holding 5 feet of water, though. Its the same with a creek. The flow of the stream is not a very significant force. It is the depth we are concerned with in making a dam.

FIGURE 5-7 shows the downstream side of the dam with more of the backup mass in place. The stones and blocks seem to be stacked rather haphazardly. Actually, there is more order than meets the eye. Notice the wall gets thinner as it gets to the top. Less mass is needed near the top because the water pressure is not much without the depth. Because the creek bottom is curved, with the low part in the center, the pool of water that is contained will be deepest at the center, and gradually shallower as the pool meets the side banks. For this reason, the dam gets thinner, or lighter, as it extends from the center to the sides, as well as from top to bottom. Over the culvert, there is some extra weight to make up for the weak, or light, area where the tube replaced stones.

I don't like to make these temporary dams one stone stronger than they need to be. If you have a very gentle flow that doesn't flood, you might be tempted to take time and make a pretty dam, because it might be around awhile. The sight of water tumbling over a stone wall has a lot of aesthetic appeal. I can't worry about the looks of a dam on this creek. I know that it is just a matter of time until a tree comes rushing downstream on a wave of water like a battering ram. I have built much more official looking dams with portland cement, big capstones, and concrete side walls, but they get busted apart, same as the fast, cheap version. The version in FIG. 5-7 has been in service for two years, and has survived two minor floods. If I can tell when one is coming, I can open the culvert to let more water through the dam, which will lessen the force on the wall itself.

One of the more ambitious kids in my area has built a beauty of a stone dam, with walls 5 feet thick. The thing has lasted through a few floods, but it also has accumulated tons of silt and leaves, which don't get a chance to wash downstream. If your pool is for swimming, your dam will need to be opened up to flush out the junk once a year or more. There is some point in building a sacrifice section in the dam that is thinner and lighter than the rest of the dam. In high water, this section will blow out and relieve the pressure, sparing the rest of the dam. You might need to retrieve only a few stones to repair the damage.

The dam in this photo series took two days to build after the materials were in the creek. Flood-damage repair amounted to four hours work for the year. Twenty hours of work a year will maintain a swimming hole in the creek. That is less than the amount of vacuum work you'd do to maintain a real swimming pool.

Once the wall is in place, with adequate backup weight, you can cover the culvert, if you have one. This alone might impede the flow enough to start filling the pool, but chances are there will be many faucet-sized leaks in the wall and under it, and you will need to put plastic over the stone or block. Sometimes when you spread a roll of plastic across a stream, it will shut off the flow almost completely. Looking downstream, you will behold a drying creek bed. If this happens, make a leak somewhere in the dam to keep water flowing downstream. When the pool is full and flowing over the top, you can patch the hole, which will put more water over the top and more air in the water.

The bottom edge of the plastic is usually forced against the dam with pressure, but in the shallow areas, where there is less pressure, you might need to put little stones on the plastic to hold it down. If your pool doesn't fill in a day feel along the bottom edge of the dam for leaks. Our creek normally flows slightly less than 1 cubic foot per second. The pool fills in one to four hours, depending on how much water is let through as it is filling. The amount of time it takes your impoundment to fill will depend on many factors. You can get a guess of the flow of the creek if there is a place along it where all the water is in one short falls. At that point, you can put a bucket under the flow and time how long it takes to fill 5 gallons. If it takes two-thirds of a second, you have a flow of 1 cubic foot per second. If it takes one second, you have a flow of 0.666 cubic feet per second.

After calculating the flow in cubic feet per second, calculate the volume of the pool you will be creating. You will only get a very rough idea, but that is good enough if you multiply the average width by the average depth by the length of the pool in feet. To find the length, sight along a level or scope set at the height of the dam to the point upstream that is level with it. The creek will back up that far if your dam works. The depth will likely be an average between zero and the height of the dam. A 3-foot dam in a shallow creek might make a pool with an average depth of 18 inches—unless the sides are sloped, in which case the pool would only average that depth in the center of the stream. Toward the sides, it would get much shallower.

You can account for the difference in depth by subtracting feet from the width measurement where the banks are steep, or by allowing a lower average depth figure. If your pool averages 1 foot deep, 10 feet wide, and 100 feet long, you have 1,000 cubic feet to fill. If your creek is flowing at about one cubic foot per second, it will take 1,000 seconds to fill if you stop all the water. If you let half of the flow go by, it will take about 2,000 seconds to fill, or one-half hour. Naturally, there is a lot of guesswork here, and your figures will not be as convenient, but I think you can guess this stuff within a factor of three, so you should at least know if your pool should fill in an hour or a day.

I find it best to fill a pool (close off the culvert opening) when the flow is good. When the creek is low, impeding the flow can really put a strain on the action downstream. If the water is high, but not flooding, you can pull a lot of the flow to your pool without hurting anything below, and the pool will fill quickly.

While your pool is filling, the water's surface will gather pollen, leaves, twigs, and whatever else normally floats down a stream. Don't be alarmed by the scummy appearance; that stuff will quickly flow out of the pool as soon as the water level spills over the dam. Occasionally, the crud will get caught on the top of the dam and you will need to toss it over. An ambitious organic gardener could skim an incredible amount of silt and leaves from a makeshift dam.

If we leave the culvert in our dam closed in the fall, over 3 tons of wet leaves will float down the stream, get stuck at the dam, and sink to the bottom in the pool. This could be seen as a gift or a nuisance. The creek also delivers us firewood. Occasional branches that fall in upstream or trees that lean too far, end up floating against the dam they can't get over. I pull them out and leave them in a pile on dry ground, where they can dry for firewood. At different times of the year, the creek brings sand, topsoil, gravel, and beer cans. I can make use of all of it, although I try not to add any of my own junk.

THE WATERFALL

Where the water finally spills over your dam is of special interest. One of the common things to go wrong on a homemade dam is that the waterfall will gradually dig a hole in the gravelly creek bottom, and the dam base will start to fall into that hole. For this reason, it is best to determine where you want the waterfall, and design the dam with that in mind. Toward the center of the dam is probably best for keeping the surface crud moving over the dam.

It is good if the water spills in a strong column into another pool of water. This will allow the fish to handle a journey over the top, and it will do the most for aerating the water. If the waterfall is spilling into a little pool, the water will act as a shock absorber, and the water will be busy making bubbles, not tearing down the dam. The pool where the waterfall drops should be lined with large stones so it stays intact. A short length of pipe at least 6 inches in diameter might be helpful in getting the waterfall to drop away from the stones at the downstream base of the dam.

Playing with the waterfalls is so much fun, you might find yourself fiddling with the top of the dam a lot, trying to get the water to drop just right. On my first dam, which I foolishly squandered too much time and materials on, the top was so level that the water would go over in a ribbon falls. I thought the thing looked great, but it didn't aerate the water as much as one pounding column in a pool would have.

FIGURE 5-8 shows an all-stone dam built between concrete bridge abutments. They made for an easy job of tying into the sides of the stream. Usually, that part of the job demands some shovel work because you need to cut steps into the banks for the rocks to fit, or you will have big leaks where the flat rock meets curved creek bottom. The straight concrete walls

Fig. 5-8. A 3-foot-high stone dam built between bridge abutments.

solved that problem. Having the dam directly under this private bridge makes for a very convenient spot to stock fish, to catch them. This dam lasted three years, until a hard flood brought a 4-foot-round tree down the creek like a battering ram, which took out the 3-foot-thick stone wall. Some of those rocks have traveled 100 feet since then. It is amazing to see the progress of something as static as a stone.

One of the oddest dam failures I've experienced is when an inner tube got left in the water overnight, and it floated up against a runoff chute at the dam top. It successfully held back the water, allowing the pool to get 6 inches deeper than it was designed for, and the extra pressure blew the dam apart. A log floating down the steam can cause the same mishap. Once the waterway is blocked partially, more stuff is drawn to the remaining leaks, and they tend to get clogged. In a way, homemade stone dams are self-sealing. A layer of dead leaves against the bottom edge of the dam, for example, will go a long way to seal up small leaks there.

It is funny how the water, when it gets too deep for the strength of the dam, will blow through the bottom instead of knocking a top stone out of the way. The paradox is that, even as the dam is about to fail, there isn't too much pressure for the top of the dam. I have had a dozen dams go down right in front of me while experimenting with depth and thickness. It is always rather dramatic and sudden, with the entire wall (which is somewhat bonded by the plastic sheet) going over, followed by a wave of water that you could surf down the creek on. My findings are echoed in Chapter 4 concerning retaining walls. The problem with walls failing is often not a case of a too-thin wall, but rather one of erosion, or poor footing. Erosion acts quickly in a stream.

Without the problems of inadequate footing, erosion, or the missing factor of the force of the creek's flow, it is safe to say that a 3-foot-thick, dry-laid stone wall that tapers inward as it gets to the top will adequately retain a 3-foot depth of water. A 4-foot depth is amazingly harder to retain and, to me, is a case of diminishing returns. Three feet of water is plenty to fish, swim, and lollygag in a raft, and you can have that much with a fairly devil-may-care dam. The extra foot of height (or depth) could double the work of the dam.

The slipperiness of stones is a big factor in the stability of the dry wall. (It is odd to call a stone wall that is almost completely submerged a dry wall, but it is.) If the stones are roundish and covered with algae, they will slide off one another with the slightest shove. Clean stones with flat surfaces are best for the amount of friction they supply. Old cement blocks are good for friction too. The weight of the wall is not always enough, if there isn't good frictional adhesion. Even a cut stone wall would push apart easily if each stone were coated in grease before being laid. In most stonemasonry, the friction between stones isn't an issue, but for a dam, where many of the stones you use will come out of the water, slipperiness can be a problem.

Another problem you might encounter on a project of this nature is a big leak that springs at the base of the dam. The pressure will push more and more water and gravel through the steadily growing hole. If you happen to be there when it happens, you'll notice no or little water going over the falls and a lot of water in a different area, which might be muddy. This is apt to happen when you build on a sandy or gravelly bottom. Pressure forces the small pieces under the dam. It is best to build on bedrock to avoid this problem, although you can fix the leak by stuffing a big stone and some little ones in the hole and lowering the depth a few inches. This erosion under the dam is much likelier with a 4-foot wall than a 3-foot one.

The less water that leaks through your dam, the more will go over the top. A falls carrying a good volume will add a great sound to the area, as well as aerate the water.

The biggest problem with making a swimming hole on a small creek is that the water volume is lowest at the prime time for swimming. If your flow is too little, the leaks in the dam will equal the flow of the creek. You will lose the waterfall and the full depth of the pool, and the surface of the water will start collecting leaves and debris. An earnest attempt at patching the leaks could reverse the trend. If you can't stop enough water to make a falls, it is best to let the creek go back to its normal level. If the flow is marginal and there is a drought, you might be losing too much water into the banks of the creek, and more through evaporation. Let the creek flow free if this is the case.

I know some people that build rubble dams in creeks just for the fun of it. I do it mostly so the kids have a place to cool off in the summer (and me too). If the creek wasn't fit to swim in, I suppose I'd find another reason to make a little dam—maybe to raise fish, maybe to skate, maybe to irrigate a melon patch, maybe for power.

If you have any fantasies about getting power from a creek with the help of a rubble stone dam, you need to make a quick assessment of the potential. If you are limited to a 3-foot-high dam (and you are pretty much using the methods described), then you are also limited to a 3-foot drop in the water, or less.

The power available in the falling water is a factor of its height and its quantity. The quantity is usually referred to in cubic feet per second. If you already have a dam and a waterfall that, through your manipulations, is dropping cleanly off the wall in one bundle without hitting anything on the way, you can measure the flow easily. It is also more realistic to measure the flow after you have a dam because then you won't count on the energy that will be leaking under the dam. Stick a 5-gallon bucket under the falls and time the filling. One and one half buckets equals 1 cubic foot. Translate your findings to cubic feet. Divide the cubic feet by the seconds to get the cubic feet per second.

The total available vertical drop of the water is called the *head*. All of the dams shown here have 3 feet or less head. The amount of power available, in horsepower, can be roughly calculated with this formula: Minimum flow (cuft/sec) × gross head ÷ 8.8 = horsepower. Other factors that should be included in the calculation are turbine efficiency and friction. When you are done figuring all the losses the system will suffer by the time you get to use the power, you will know that either you need to have a passion for water power or have no alternative power in the area, to bother with the amount of electricity available in a small-volume, low-head situation. I calculated the power I could tap from my dam setup once, and it was on the order of 100 watts. If I had managed to intercept that force of falling water, the bubbles wouldn't get made, and I'd need to put a 100-watt aerator in the creek. So I decided to leave it alone.

The fastest way to tap meaningful electricity from a small waterfall is to set a back wheel of a bicycle with a bike generator under the flow. It should spin the wheel, which should spin the generator, which should light the light. This is an alternative energy project within the grasp of most people in the United States.

Currently, most alternative energy in the United States is on the level of a tinkering hobby. Water power is no exception. (Water power is solar power with a time delay.) If you get into it, it should be because you want to check switches, gauges, and levers; add fuses and meters; and go to hardware stores, where people will think you're nuts when they ask you why you want the obscure adapter they don't have. It shouldn't be something you get into because you have a vision of free electricity. The electricity that comes from the big grid and costs less than a dime for a kilowatt hour is free, or as close to free as you can get.

Gardening is an apt analogy. Nobody in this country would grow potatoes because they cost too much in the store. They don't cost much in the store, and anybody who grows them knows that. The reasons for growing your own food are to do something meaningful, to have a relationship with the earth that is comprehensible, to have a pleasant hobby, to get better food without toxins. It is the same with making your own power. Don't try to make economic sense of it; it is a spiritual issue.

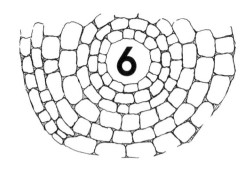

6

MORTAR

AFTER the discovery that stones could be stacked over one another into shelters and other structures, it was inevitable that people would want to fill all the gaps in the dry-laid stone—to hold in heat, to keep out bugs, to block the wind and rain, and to create chimneys that would contain and direct smoke. Clay was probably the first mortar. People found that they could make a stiff mud to bed stones and it would fill the gaps between stones, while setting firm enough to support the weight of several feet of stone. The problem with clay mud is that it will erode out of the wall fairly quickly. There is constant maintenance. Lots of things were done to clay to make it last longer: sand was mixed in, horsehair and other fibers were added, ashes were stirred in. It was not until people started investigating natural cements—that is, the substances that hold sedimentary stone together—however, that much progress was made in the search for the perfect mortar.

LIME AND CEMENT

If you look closely at a piece of sandstone, you can see that it is composed of grains of sand cemented together. The cement is hard to see, but it is there. In pure sandstone, the cement is a very fine siliceous material. Pressure and water are also needed to cement the tiny stones called sand. Siliceous cement is the strongest, but very hard to imitate. The big jump in the development of masonry came when people figured out how to take apart limestone, and basically put it back together again. Perhaps someone found some burned limestone after a horrible fire and started the experiment. Who knows how it was discovered or when.

Pure limestone is calcium carbonate ($CaCO_3$). Common impurities are silica (SiO_2), alumina (AlO_3,

and iron oxide (Fe_2O_3). If the limestone has less than 10 percent of these impurities, it will make quicklime.

Quicklime

When limestone is heated to about 850°F, called *calcining,* it gives up carbon dioxide and becomes CaO. Pulverized, the calcium oxide is called quicklime. When water is added to quicklime, called *slaking lime,* heat is given off and the quicklime takes on the hydrogen from the water (and gives off oxygen) and becomes calcium hydroxide (CaOH).

Slaked Lime

Slaked lime is quicklime that has reacted with water, but is kept away from the air. Quicklime will react with a volume of water one-third the size of the lime. The water allows the quicklime to become CaOH. This slaked lime is either used as a putty, or allowed to dry and used as a powder. When this powder is exposed to air, as in mortar, it gradually takes on carbon dioxide and becomes limestone, once again. Slaked lime today is called *hydrated lime.* If you buy a bag of it, you will instantly see that it is a much expanded, or fluffed-up product compared to agricultural lime, which is crushed limestone. FIGURE 6-1 shows how mason's lime is made.

Hydraulic Cement

When limestone with the right mineral impurities (magnesium and others) is processed for cement, it gives the mortar the property of setting underwater. Hydraulic cement was used by the ancient Romans over 2,000 years ago. The original mortar in some of the ruins is still in good shape.

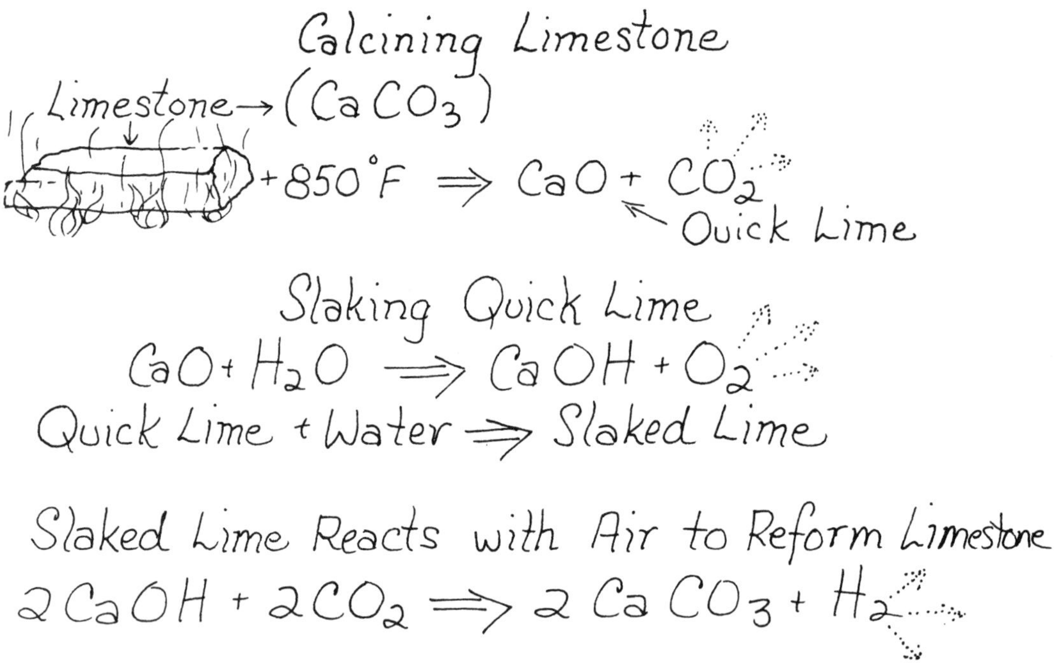

Calcining Limestone

Limestone → $(CaCO_3)$

$+850°F \Rightarrow CaO + CO_2$

Quick Lime

Slaking Quick Lime

$CaO + H_2O \Rightarrow CaOH + O_2$

Quick Lime + Water $\Rightarrow$ Slaked Lime

Slaked Lime Reacts with Air to Reform Limestone

$2CaOH + 2CO_2 \Rightarrow 2CaCO_3 + H_2$

Fig. 6-1. The journey of limestone in mortar.

Pozzolan

The early hydraulic cement of the Romans was called pozzolan, named after a city near a volcano that spewed up a certain mineral, called pozzolan, which was mainly silica and alumina. It was ground up and added to the lime, and that mix was added to sand. In this ancient mix, much more cement than sand was used in the mortar. Pozzolan today refers to a portland cement that has an artificially ground pozzolan added for the special property of setting without giving off much heat. It is softer than straight portland cement.

Lime-Sand Mortar

Sand and slaked lime alone would make a pretty good mortar, it was discovered, and that is mostly what was used for masonry until the advent of portland cement. Sand-lime mortars are seldom used today, but they still exist in most stone structures built before 1850. The lime cements the sand particles together into a sort of calcareous sandstone. I must say, stonemasonry has not advanced since the discovery

of slaked lime and sand mortar. The newer cements have allowed the invention of concrete, which has understandably taken the place of most stonework.

Portland Cement

The discovery of portland cement was a slow, inevitable process. Basically, the same technologies were employed as those used for making lime—mining, kilns, and crushing devices. Supposedly, portland cement was first made in 1827, in England. it was fairly slow to catch on. By 1897, only one-third of the cement being used in the United States was portland, but by 1907, 90 percent of the cement was portland.

Portland cement makes a mortar that is much denser and stronger than natural cement, and this mortar has allowed all manner of changes in masonry. Oddly, the advent of the near perfect cement has spelled near death for the craft of stonemasonry.

Portland cement is a mixture of calcined limestone, clay, and gypsum ($CaSO_4$) in just the right proportions. The burnt mixture is then pulverized to an incredibly fine powder. Portland cement particles will go through a sieve with 40,000 openings per square inch. The powder is then bagged in cubic-foot quantities, which weigh about 94 pounds. Some of the advantages and disadvantages of portland cement will be illustrated throughout this book.

Natural Cement

Natural cement is a calcined argillaceous limestone, or essentially, lime with fireclay. Natural cement weighs less than portland, and sets weaker and faster. When mixed properly and cured well, natural cement can have advantages for stonemasonry in which great compressive strength isn't needed. Natural cement is still available in some outlets and has been used in this country since 1818. Portland cement is an artificial combination of minerals to create an ideal mixture, whereas natural cement is a processed limestone chosen for its impurities.

AGGREGATE

In a mortar mix, the cement is called the *binder.* A binder is usually stretched, or better utilized, with the addition of aggregates. The most common aggregate for mortar is sand.

Sand

As the aggregate in cement mortars, sand can profoundly affect the finished product. The amount and kind of sand in a mortar mix have as much to do with the final product as does the kind of cement used. The size and shape of the little grains have a lot to do with the way they behave in a mortar.

Sand starts out as magma, below the earth's crust. The magma comes out in a volcano and hardens into a basaltlike stone of mostly silica. Eons of weathering wear the mother stone into tiny pieces, which erode down slopes via streams and rivers until they reach the ocean. There the little pieces accumulate and build up beaches. The sand particles build up deeper and deeper until there is considerable pressure at the bottom of the pile. The pressure fuses the grains into sandstone. At some point, the earth's plates collide and slowly force up mountain-sized chunks of the crust, carrying sandstone along.

There is sandstone on the mountaintops near my home, and when it rains, little pieces of those stones start to work their way down the creek to the bay, where they are building up in a layer deep enough to start creating more sandstone.

For use in mortar mixes, sand should be free of organic matter, which would create weak spaces in the mortar. In a lime-sand mortar, it has been shown that as much as 7 percent clay powder in the mix will actually strengthen the mortar. Natural sand, or *bar sand*—sand that accumulates on river bars—is composed of grains of many different sizes. Because of the variety of sizes, there are fewer spaces (about 30 percent) in natural sand than in mined and sifted sand (over 40 percent spaces). Masons often avoid natural sand because its rounded edges do not offer as much adhesion as do the sharp grains of mined sand. Yet with its lesser voids, round sand can be used well in stonemasonry, especially if more cement is used to compensate for the lesser adhesion.

Coarse sand has less surface area than fine sand, and therefore requires less cement/water paste to be coated. This means you can make a stronger mix with coarse sand than with fine sand, using the same amount of cement. I have seen 150-year-old pointing mortar of round bar sand and lime that was still in great shape. The thin joints had surprisingly large grains of sand.

The strongest, densest mortar is made with sand that is 78 percent coarse grains (larger than 1/16 inch) and 22 percent fine grains (smaller than 1/50 inch). In this arrangement, the fine grains fill the spaces between the big grains. Fine sand alone, without coarse grains, is 44 percent spaces. Coarse grains alone are only 37 percent voids. Perfect spheres of the same size, packed as tightly as possible without squashing, will have 26 percent spaces. The idea with strong mortars is to fill all these hollow places. Mortar has a much greater compressive strength when it doesn't have a lot of empty spaces.

Masons often opt for fine, sifted sand, in spite of its need for more cement to bring up the density, to achieve good strength. The fine grains make for a very workable mortar and a very fine finish. Big grains are clunky to the trowel. This is far more significant in brick and block work than in rubble stonework. With rubble, you'll use so much mortar to fill all the spaces between poorly fitting stones that you might as well save money and use coarse sand.

When a mix of one part portland cement and three parts coarse sand (containing 22 percent fine grains) is made in laboratory conditions, the tensile strength of the mortar achieves 700 psi. The same mix, using all fine grains, obtains only 189 psi, so the difference is significant.

Before you worry that your cement won't be strong enough, remind yourself of the European castles and cathedrals of stone, laid in lime mortar. The stresses on the stone and mortar in a house or barn are minimal, making the quest for mortars of ever increasing tensile and compressive strength rather absurd. The heaviest load any stonemasonry on earth is subjected to is quite possibly the granite piers of the Brooklyn Bridge towers, which at the bottom stone carry about 400 psi.

As it turns out, the way you lay the stones has a lot more to do with the final strength of your wall than how you made the mortar. A well-laid stone wall could go 60 feet high on the weakest mortar any modern mason would be caught using. A poorly laid wall will fail no matter how good the mortar is.

You can test the density of your sand—that is, check how much empty space is in your sand—by seeing how much water will fit in a 5-gallon bucket of dry sand. If the sand will hold 2 1/2 gallons of water, it is half spaces. If it holds only 1 gallon, it is 20 percent voids, and so on.

You can also see how much dirt is in your sand by shaking a sample in a jar of water. The lighter silt and junk will settle on the top of the sand. Usually, you can tell by looking if the sand is too funky. Any respectable masonry supply place will sell clean, sharp sand, usually in several grades. Finer sands lean toward whiter colors and cost quite a bit more.

A cubic foot of sand can vary a lot in weight, depending on the percentage of voids. Dry sand can be 86 to 123 pounds per cubic foot. Sand of 123 pounds per cubic foot is obtained with an artificial mix of properly proportioned grains. Sand heavier than that can be obtained with pressure: squeezing the pieces into a tighter arrangement, as happens with sandstone, which weighs more than 150 pounds per cubic foot. In theory, stronger mortars could be made with heavy metal grains, because they are heavier than sand particles. For economic reasons, sand is the mortar aggregate of choice.

Other Sand

The chemical composition of sand is not a very significant factor in making mortar. The grains of sand don't enter into the reaction; they merely get stuck in it. When you crush old mortar, the sand is still there. Grains of other materials will also work as fine aggregate in a mortar mix. The screenings, or tailings, in the manufacturing of limestone gravel make a very strong and cheap mortar. If you live near a quarry, these tailings cost only a few dollars per ton. They make an exceptionally strong mortar because of the wide variety of grain sizes. Some of the pieces are 1/8 inch in diameter, and there is fine dust as well. It makes a very rough mortar, but the cheap lime sand will go twice as far with a bag of portland cement as regular sand will. I see no reason not to use it—except as a finish mortar for fancy work with tight-fitting joints because it has such a rough texture and ugly, gray color.

I have seen surprisingly strong mortar made with sawdust as an aggregate. One wonders how long the little pieces of wood will be stable. I heartily encourage experimentation along these lines, yet I would be foolish to advise the use of unproven techniques. Perlite is often used in concrete to lighten its mass. It acts as an aggregate, but severely weakens the concrete. Mortar can be made with added soil, to stretch the cement dollar. I have seen and worked with soil cement that is now 10 years old and doing fine. The mortar is hard and not cracked, even though it has more dirt than sand or cement. I can't say I recommend it, although I don't doubt for a second that someone could build a lasting stone structure laid in mortar stretched with clay soil. Again, the more significant factor is how you fit your stones.

Gravel

Gravel is coarse aggregate used in masonry, though not often stonemasonry. To me, gravel is any stone larger than 1/4 inch cube. Smaller stones are coarse sand, which is lumped under the fine aggregate category. Most of the gravel—or *stone,* as it is often called—that is sold for home concrete use has a maximum size of 3/4 inch. *Pea gravel* is not a technical term, but is often used to describe gravel with grains slightly larger than sand, up to about 1/2 inch.

Gravel is like sand in most ways. It has a large percentage of voids unless it is graded, which means it has pieces of different sizes to minimize the voids, and maximize the effectiveness of the cement.

Angular gravel offers greater tensile and compressive strength than rounded, river-bar gravel. Most masonry books reject its use, but for around the house, round gravel will make a hard enough concrete for almost anything you will need. If you happen to have round sand and gravel on your place, go ahead and use it. Add a little extra cement.

I have used gravel in mortar mixes for laying stones with good results. The gravel makes a batch of mortar twice as big without raising the cost much. Working with concrete as mortar is rough and heavy, but it makes for fast work. You can fill much larger spaces between big stones with concrete before it will slump out. Mortar will spill out of the wall if it is in too big a blob. Putting a shovelful of concrete between faces on a course is so much faster than doing it trowel by trowel with mortar, looking for little stones to break up the big lumps of mortar.

Concrete can handle a lot more weight than mortar before it will squish out of the wall and run down the stones. If you try to stack stones 8 feet tall in one column in a day, the mortar at the bottom will get wrecked by the weight before it has time to set. You can't really stack much higher than 2 feet in a day without messing the mud, except with gravel mortar. You can use a mortar mix later to point stone laid in concrete.

Following is how I mix concrete for use as a bedding mortar. Using the biggest wheelbarrow they make, mix three shovels of portland cement with nine shovels of sand and enough water to make a runny mix. Then start adding shovels of gravel and mixing them into the stew until you can't add any more and still get all the stones coated in the cement paste. You might fit 18 shovels in there, which will double the size of your load.

Gravel mortar is best used to fill cavities in walls, although you can lay stones in it, if you maintain space at the face of the wall for a pointing job later. Generally, a mortar joint shouldn't have an aggregate more than one-third the size of the joint. To keep an integral mass of concrete or mortar 1 inch thick, you would need to use aggregate no larger than 3/8 inch. The smaller gravel, called *1B stone* where I live, is probably best for inclusion in a mortar.

The usual gravel sold for concrete work has 30 percent voids. Using graded sand with gravel will reduce those spaces. The cement particles fit in the spaces between fine sand grains; the fine sand grains nestle between large grains of sand. This mixture fits in the spaces between gravel stones. The procedure can continue to include large stones, although gravity begins to play a more important role with big pieces. If they are positioned to fall, they will overcome the adhesive strength of the mortar.

MORTAR

Mortar got its name from the vessel in which it was mixed. It is a semiliquid mixture of sand and cement, as aggregate and binder, with enough water to make a paste. Mortars made with one part portland cement and three parts ideal sand can obtain compressive strengths of 8,000 psi! This is strength beyond anything anyone needs in masonry, yet it remains a popular mix. When people hear the word *strength,* they figure that a strong mortar will bail them out of not understanding masonry. The problem is, with all its compressive strength, the portland mortar is not a very good glue, especially if it is holding something against the will of gravity or if the stones are dusty before they are glued.

Note that figures about strengths of different mortar and concrete mixes so frequently quoted in building textbooks are for ideal laboratory conditions. The perfect ingredients are kept at 70°F and moist for 28 days. This is never the case in real life. Don't get snowed by meaningless claims about strength. Weight, despite all the attention cement gets, is the real strength of masonry, especially rubble stonemasonry. If you could perfectly glue a wall of stones into a single unbreakable mass, the wall would still only be as strong as its weight. The best that mortar can do for a hollow-core, 8-inch block wall is to make it a single unit. Even if that was possible, the block wall, with its 8,000 psi cement mortar, would be eight times easier to push over than an old rubble-stone wall 2 feet thick and laid in dirt.

Again, portland cement is not necessary for good stone or brick masonry. A mixture of 1 part lime to 4 1/2 parts sand will give a mortar that can handle 100 psi, and that will do. A mix of 1 part lime and 3 parts sand will nearly double that strength. This seems to be the mix favored by old stonemasons, although you can find accounts of buildings being laid with a 1-to-1 mix. Walls built with lime and sand mortar have stood for 800 years, and need repointing about once every 200 years. This amazing track record is more to the credit of the technique of stacking than to the type of mortar.

Using portland cement more than doubles the strength of a mortar and also adds the convenience of a high early strength, which lets you build without wrecking the mortar from the previous day's work. The problem with portland cement is that it contracts slightly as it sets, causing cracking and crazing. Also, the mortar approaches or exceeds the strength of the units, thereby losing its value as a sacrifice, or shock absorber. Bricks laid in portland cement mortar will often break if there is any motion in the building, while the mortar joints will remain unharmed. The correct approach is for the mortar to give during stress, and spare the brick or stone.

Portland cement mortar can be *retempered*—that is, remixed after the initial reaction has begun—to reduce the amount of shrinkage during setting, but it will also reduce the strength by as much as 50 percent. Retempering also leaves mortar more plastic for a longer period of time. In some cases, the loss of strength is worth accepting for the greater workability of a retempered mortar. Most masons I know do it, although most manuals frown on the practice.

There are so many types of cement, and so much testing has been done, that masonry books abound with facts and figures about mortar and concrete. I would like to make these blanket statements on the subject, in the hopes of making the whole subject more comprehensible to the handyman: The very best masonry on earth, which will probably outlast all the competition, is built without mortar. To defend this statement, I point to the flawless mortarless walls at the Lost City of the Incas at Machu Picchu, the great pyramids of Egypt, and the domed trullo houses of Cisternino.

In the next category are structures like the Taj Mahal, the huge cathedrals of Europe, the Scottish castles, and the Great Wall of China. These were all built with mortar, but without Portland cement and all of its great attributes.

The advent of portland hasn't helped stonemasonry at large. Instead, it has tempted people into making unsound stone structures that are undersized and nonintegral. A stone wall made in a form with portland cement is like a piece of plywood. It is a new arrangement, dependent on glue more than time-honored fitting. What happens when the glue fails? Gravity is unfailing glue.

The temptation with portland is to ask it to do impossible things, like prevent a wall that has an inadequate footing from falling over, or keep water from coming through a basement wall. The super-hard mortar from a rich portland mix is prone to cracks, so even though it is so hard, it still falls out of stone walls and lets moisture into them. The hardness is not really needed. On repair jobs, I continually find walls with big patches of portland mortar. The patch never solves the basic problem with the wall, but it sure makes a hard mess to get off the wall.

There are many types of cement, and mortars come in types, too. For any work around the home, three products will get you through any mortar needs: sand, of the regular to coarse variety sold at concrete yards; type 1 portland cement, which comes in a 94-pound bag; and type S hydrated lime, which comes in a 50-pound bag. By adjusting the proportions of these three ingredients and water, you can make all manner of mortar mixes for every possible purpose. Sure, there are cements for high heat or acid resistance, but all you need are sand, lime, and portland cement—unless you want to color your mix with pigments, in which case you also need pigments and type 1 white portland cement instead of the gray.

White portland cement allows the other colors to show. White cement is often used in restorations because it gives the same look as the old lime mortars.

Mortar with mostly lime and sand is slow to reach its final strength, and needs diligent curing water so that it doesn't lose its water to evaporation before it has had a chance to become part of the new material. (See Chapter 12.)

Air-Entraining

Air-entraining is the process of adding an agent to cement or lime, causing it to make tiny air bubbles when it is mixed. The tiny bubbles get trapped in the mortar or concrete, and are evenly dispersed and similar in size. The purpose of the tiny air bubbles is to help concrete or mortar get through freeze/thaw cycles without cracking and breaking. The tiny spaces give the concrete somewhere to expand to without busting apart—like miniature shock absorbers. The lime you buy will probably have some air-entraining agents in it, since it is a desirable feature for almost all types of modern masonry.

Masonry Cement

The substance called masonry cement, which is sold at masonry supply outlets and comes in a cubic-foot bag weighing from 69 to 72 pounds, is a mixture of half hydrated lime and half type 1 portland cement by volume. This is the all-purpose stuff that is normally used with three parts of mason's sand for laying brick and block. The mortar it makes at this ratio is called *type N* mortar.

If you want the least confusion, all you need for stonemasonry are masonry cement and sand. To bed stones, you can mix one part cement to four parts sand. For pointing, mix the mortar one to three. For extra thin joints, where needing extra adhesion or when using fine sand, mix one part cement with two parts sand. Masonry cement also comes in white, at about twice the price, or black, at three times the cost of regular. There are other colored cements available.

The problem with masonry cement, or the pre-mix of portland and lime, is that you can't fine-tune the ratio of lime and cement, only the ratio of sand and cement. It is like buying ice tea mix that already has lemon and sugar—sure it's convenient, and you can make it strong or weak by adjusting the water, but what if you want less sugar and more lemon? In masonry, you can also buy a mortar mix with the sand already in it. This is very easy and very expensive—like a T.V. dinner. The proportions are set. If you like more peas and less apple brown Betty, you're out of luck. If you're doing anything but the tiniest of masonry jobs, don't get anything premixed. If you buy lime, portland, and sand, you'll be able to fine-tune your possibilities.

Another common mortar type is *type M*. It has greater compressive strength than type N and greater durability in contact with the earth. It is often used in foundations, retaining walls, and sewers. It is made with one part portland cement and one part masonry cement to six parts sand. If all you have is portland cement and lime, you can put together type M mortar with three shovels of portland cement and one shovel of lime to twelve shovels of sand.

Type S mortar is used when extra adhesion is needed. If you are thinking of gluing a stone in place, you would want to use type S mortar, which can be made with two parts portland cement, one part lime, and nine parts sand. Type N, which is regular mortar, if there be such a thing, is best when the masonry is exposed to harsh weather. It can be made with one part portland cement, one part lime, and six parts sand. You can see that buying both lime and portland cement allows you all these variations, which you can't have with a bag of regular masonry cement.

Again, mortar for thick, laid stone walls that aren't extraordinarily tall doesn't need great compressive strength. If it did, you wouldn't be able to lay stones. Think of this: The 2 feet in height you lay up in a day is being supported by a mortar that is still semiliquid. If that soft stuff will support a few feet, how hard does the mortar have to be to support 10 feet? The average forces are easy enough to calculate if the wall supports only its own weight. A cubic foot of masonry weighs about 144 pounds and exerts 1 psi on the ground. A 10-foot-high wall presses the ground or footing with an unincredible 10 psi. Walls that carry a roof load might have twice that weight. In actuality, certain square inches will support nothing, while other spots might have triple the average weight. Still, nothing on the order of 3,000 psi will ever occur around the home.

The best example I can think of to illustrate the role of mortar in stonemasonry is a set of building codes that were established in Chicago near the turn of the century, to set safe limits for different types of stonemasonry. These guidelines were developed by the leading architects and engineers of the day to determine how much weight could safely be handled by different structures. These are not breaking strengths:

☐ For a structure of uncoursed random rubble stone laid in lime mortar, they allowed 60 pounds per square inch.
☐ For an uncoursed, rubble structure laid in portland cement mortar, they allowed 100 psi, quite a jump in strength.

☐ For a *coursed* rubble wall laid in *lime* mortar, they allowed 120 psi.

The implication of these figures is that you get more of an overall strength increase by switching from an uncoursed to a coursed method, than by changing from lime to portland cement. According to those codes, there is another doubling in strength when you switch from rubble to ashlar masonry. This implies that the shape of the stones and the way they are stacked are more significant for strength than the kind of mortar used. I certainly agree.

MIXING MORTAR

To mix your own mortar, you'll need either a portable motorized mixer, a mud box, or a wheelbarrow. If you are working by yourself, especially if the project is small, you won't be able to use up mortar fast enough to justify the hassle and expense of a gas or electric mixer. If you had a crew of three, a mixer would be a big help, and you could use a whole bag of cement at one time.

If you decide to mix by hand, I recommend a big, strong wheelbarrow; a big, strong hoe; and a big, strong person. Naturally, all the size can be scaled down, but at whatever scale you use, the tools need to be strong to push the heavy ingredients around, and the container also must be heavy duty. Mixing cement in a little lawn wheelbarrow with a garden hoe can leave you with broken tools and not a lot of mortar. It is nice to have a container larger than the batch you mix to minimize the slopping. If you mix a big batch, you need a stout wheelbarrow to move it around.

I use a wheelbarrow because I can mobilize the mortar and put it right where I need it without much shoveling. It is also helpful to move the batch to various stations where you have the sand pile, the cement, the lime, and the water. If you have sand delivered, which makes sense on all but the smallest jobs, the dump truck might not be able to get next to the building site. With a wheelbarrow, you can shovel the sand in at the sand heap, then push the load to the shed where you have the cement and lime. It is handy to have the water there, too, at the end of a hose that can be shut off at the mixing site.

The normal alternative to the wheelbarrow is the mud box, which is a big pan with a flat bottom and angled ends. A mud box is often built on the job from plywood and 2- x -6 lumber scrap. You can mix a big batch in the box, but you will have to shovel it out of the box and move it in a wheelbarrow. Why not start in the wheelbarrow if you work by yourself, since you won't be able to use more than a wheelbarrow can hold before the mortar gets stiff.

I lay stones in distinct stages, with different mortars for the different phases. First is the footing, which can be made with pretty junky mortar. Next is bedding, which can be done with a weak (by today's standards) mortar that is easy to clean off stones and is strong enough for its intended purpose. Pointing mortar is last, and needs to be stronger and more adhesive.

Bedding Mortar

For bedding stones in rubble masonry that is coursed, a good mix to use consists of 1 part masonry cement and 4 1/2 parts sand, preferably coarse sand. For bedding mix that won't show, there is no reason to use expensive colored or white cement, or costly fine sand. For house-sized structures with thick-laid stone walls, rocks can and have been bedded in clay mud. As long as the clay can't get out and water can't get in and cause the clay to freeze and expand, the compressive strength is not an issue. These days, it is easier to make a weak mortar mix for bedding than to find a suitable clay. FIGURE 6-2 shows a good mix for bedding stones. The lower part of the illustration shows an equivalent mix using masonry cement instead of portland and lime.

Bedding mix moves pretty fast on a job, so make as big a batch as will fit nicely in the wheelbarrow. Try 3 shovels of masonry cement and 13 1/2 shovels of sand. That should fit with enough room to hoe. Stay in the same proportions when you change the size of your batches to accommodate a change in working pace.

Put the sand in the wheelbarrow first, and spread it out as much as possible. Next, shovel in the cement, spreading it roughly over the sand. You get this much mixing for free. Then fold the dry ingredients into each other, scraping hard at the corners of the pan to get all the sand out into the mix. When the dry mix is done, start adding water slowly while mixing it in, as though you were making batter for pancakes. Pour the water in from a 5-gallon bucket (or other big pail),

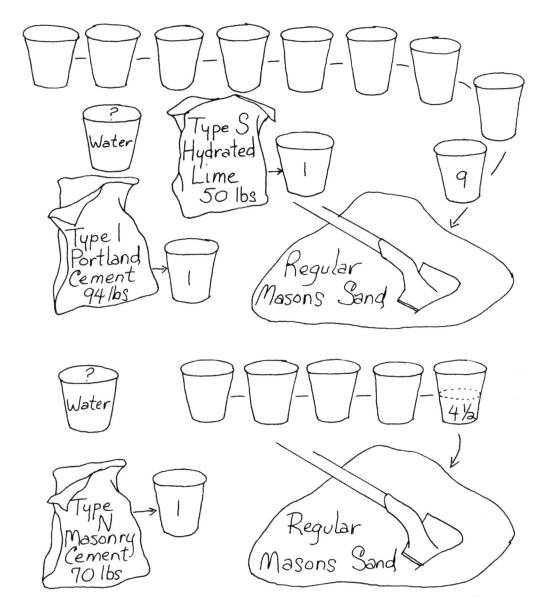

Fig. 6-2. Mortar mixes for bedding rubble stone.

and take note of the amount of water that is needed to get the mix right. On subsequent batches of the same size, you will know about how much water is needed and won't have to worry about getting the batch too wet.

The problem with mortar is almost always a case of too wet. A high slump, or dryish mortar (and concrete) is less messy and much stronger. Stonework

requires a stiffer mud than brickwork because the units are so much heavier and because the stone is much less absorbent than brick or block. A brick will suck the juice out of a mortar that is almost sloppy wet, and it will stiffen quickly. Stone won't do that unless it is exceptionally porous.

Wet sand can cause you to make a too-wet mortar. Soaked sand can hold 3 quarts of water per cubic foot, and drastically alter your mix from what it would be with dry sand. You can get your batch right again by adding more sand and cement in the proper proportions. It is harder to mix them in well because there is no way to dry-mix the ingredients, unless you use another mud pan. You can also run out of room pretty fast if you were trying to make a big batch and suddenly you need to make it bigger to get it dryer. If all of your sand is soaked, you will need a lot of sand and cement to soak up the excess water. It is a lot easier to not overwater when you first mix the mud.

Probably the hardest thing a mason must learn to do is count without losing track, sometimes as high as 16. When shoveling out the ingredients for a batch of mortar, you must not get sidetracked. It is amazing how often a grown person cannot count to 3.

I have developed a method of hand-mixing bedding mortar which dramatically reduces the amount of work required. The method is in contradiction to most information I've read, but I've used it successfully for years and the mortar has always set as though it were mixed the normal way. Here's how it works: When you have found the size batch you can handle well and have determined how much water is required to get the batch just right, you needn't do any dry-mixing of ingredients. You can shovel in the sand (spread out), then shovel in the cement . . . except for one shovelful which is used later. Next, dump in all the water. Because you are missing a shovel of cement, the mix is going to be much too wet. You can use this to your advantage. Start to mix the soupy mix and push all the dry stuff into a slurry. Pushing the wet mess around is an easy way to have the dry ingredients disperse themselves.

It is much easier to push the wet stuff around than to push and stir dry sand and cement. A kitchen analogy: If you wanted different vegetables evenly distributed in a stew, wouldn't it be easier to toss them in the pot and stir a few times, than to prestir the dry vegetables and dump them in?

When the slurry is well mixed and there is no sign of different colors or shades in the batch, you can stir in that last shovelful of cement, which will bring the batch to the right consistency. For laying stones, the needed consistency is something like peanut butter.

I am confident to recommend this work-saving approach to mixing bedding mortar because even if it isn't as perfect as mixing the dry ingredients first

and slowly adding the water, I know it is a lot stronger than the bedding mixes used in the old days, in buildings that are still in great shape. I have done demolition on old stonemasonry, and right behind the crust of lime-sand pointing mortar is often dirt. Sometimes the dirt seems to have a little lime in it and is crusty; other times the stuff spills out of the wall like powder. I've never encountered an old bedding mix that couldn't be crumbled between the fingers. That doesn't imply that the buildings were crumbling. If the wall is laid well and has a weatherproof pointing mix to keep the bedding trapped in the wall, a structure laid in clay dirt could last a very long time.

No modern masonry book would suggest laying stones in dirt for a house, and I'm not advocating it. I'm just saying you could do it and it would work. There are stone houses with dirt between rocks that will outlast the block and concrete structures being built now. It is not because of the dirt, certainly, but because of the stacked walls, 2 feet thick, of overlapping layers of squarish stones.

Pointing Mortar

When you are laying rubble stone in a weak bedding mix, you need a strong pointing mix at the face of the wall, extending into the wall about 1 inch, to keep the bedding from eroding. In the past, slaked lime and sand made the best pointing mix. All manner of variations have been used in old buildings. Mortar has always been something of an experiment, and it still is. Many old stone buildings in England are laid and pointed in mortar that had more lime than sand. Others had three times more sand than lime. It is all still debatable.

There's not a whole lot of agreement among the supposed authorities on the subject of what mortar mix to use. One thing that is seldom pointed out is that, if strength was all that mattered, you can't beat straight portland cement with water. A 1:1 ratio of sand and portland has much greater tensile strength and compressive strength than a mix with twice as much sand. Yet no one is suggesting that kind of mix, because there is a lot more to it. There is economy, flexibility, workability, waterproofness, adhesiveness, and a host of other considerations.

FIGURE 6-3 shows a good mix for pointing. If you want it to be easier to clean, use white portland or

Fig. 6-3. Mortar mixes for pointing stone or brick.

white masonry cement. The bottom of the illustration shows the same using masonry cement instead of portland and lime. This is to show ratios, not quantities.

Six buckets of sand would be much too big a batch. White cement costs about double, but it is worth it for pointing mortar because so little is needed.

When you point a wall, you're only applying a thin matrix of mortar over the spaces between stones at the surface of the wall. One bag of cement will go a long way on a pointing job. I recently pointed 120 square feet with one bag of expensive black cement. It was the entire job. I could have saved the customer $7.00 using regular cement, but it wouldn't have had the

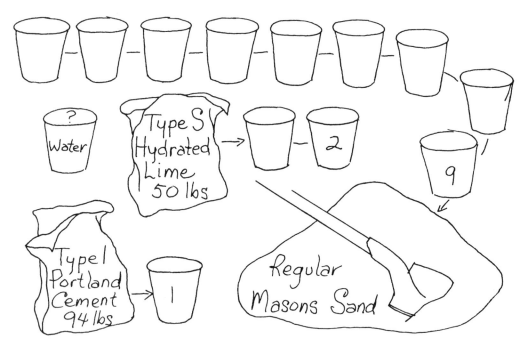

Fig. 6-4. Mortar mix that more closely resembles old, good pointing mortar.

same appeal. Bedding a wall in black mortar, just to get the surface joint black, is a waste of money.

Generally, there are some savings to be had in mixing your own masonry cements with portland cement and hydrated lime. The larger the job, the more sense it makes to do that.

When mixing a pointing batch, make sure the wheelbarrow or mud pan is free of little dried mortar chunks. These can be aggravating if they get in the mix because they won't fit well in the joints. It is also good to have clean sand for pointing. Leaves are a hassle in mortar for pointing.

The amount of pointing mortar that can be used up by one person before it gets too stiff is much less than a bedding mortar batch. You need small batches of high-quality mortar. Three shovels of masonry cement to nine shovels of sand should be enough at one time. This quantity is so easy to move around with

a hoe in the wheelbarrow that I generally dry-mix the ingredients for pointing mortar. Everything about the pointing mud is more demanding than the bedding. It must wear better, be less absorbent, have greater adhesion, look better, and so on.

There is evidence that a limy mortar has the characteristic of sealing its own little cracks and fissures as it reacts with the carbon dioxide in the atmosphere. Carbonic acid is formed, and it dissolves some of the lime, which is redeposited in the cracks. A mostly lime-sand mortar will also expand slightly on setting, giving a very tight seal around the stones. Lime improves the water resistance of a mortar, as well. FIGURE 6-4 shows a pointing mix with a greater percentage of lime, to take advantage of these properties. This would be a good mix for pointing old brick, too. More care must be taken in curing limy mortars. (See Chapter 12.)

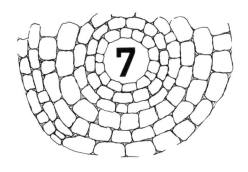

7

STONE FOUNDATIONS

BUILDINGS with wooden frames are often raised above grade on a foundation of stone. The foundation rests on a footing, which is the base for a foundation. A foundation can either be a continuous wall or a series of pillars or piers. Pillars are much faster to build than a complete foundation, and many builders opt for that shortcut in getting a roof over their heads. Afterward, when the house is done, stone can be laid in a thin, nonstructural, non-loadbearing wall between the pillars, to complete the effect of a complete stone foundation. I call that procedure *under-pinning*.

The houses used to illustrate this chapter (and others) are owner-built, small first houses. I had the good fortune to get my sweat in these houses, somewhere in the mortar. All of them are still occupied and beautiful. The log cabin seen in FIG. 7-1 is built of poplar trees from the immediate vicinity. The whole structure is kept level and off the wet ground on stone piers made from rocks taken out of the building site. The original building cost less than $1,000, and took two people (with no building experience) three months to build. It was completed during a summer, so the builders were able to camp out at the site.

STONE PIERS

When you are building on top of stone piers, each pier will have to carry more weight than if the foundation was a continuous wall. It is important to build the columns soundly, and that means starting with a good footing. In some places, the ground itself is a good enough footing, but in most places you will need to excavate all the material that will move, whether it be from freezing, mushing out, or erosion. The building must sit on something that won't move.

Footings for Stone Piers

Normally, the footings for stone columns would be very broad to evenly spread out the weight carried by the foundation. If the ground is steeply sloped, some of the piers will be much taller than others to give a level foundation. Laying rubble stone more than a few feet high requires a minimum thickness of 18 inches. For this reason, the smallest, stable stone column you could build more than 2 feet high would need to be 18 inches × 18 inches. If the stone is rough, it will need to be at least 24 inches × 24 inches. If the footing is to spread the weight borne by the column, it must be broader still. A footing 36 inches square, dug plumb on the sides down to the frost line and firm subsoil, should be fine for all but the hugest of houses on the softest of subsoils. For a shed or other single-walled, single-story building, the footing could be the same dimension as the column, since there is less weight to spread.

If you are building where there are outcroppings of bedrock at the surface, you could build directly off the bedrock with as narrow a column as you can stack well with the stones you have. The surface of the bedrock would need to be pretty level where the column sat. If you are building on a level surface, your stone piers need only be a foot or less off the ground, if the purpose is merely to keep a wooden structure out of contact with the moist ground. If the columns are that short, you certainly wouldn't need to build a 24-inch square one; instead, you could use a single stone that sat well on the footing and had a flat top to support the wooden beam above it. If a stone came to a point, naturally it would not support a wooden beam or sill log well. The top surface of a single stone pier should be at least as wide as the width of the wooden beam, as shown in FIG. 7-2, (LEFT).

Fig. 7-1. A log cabin built on stone piers.

Right

Wrong

Fig. 7-2. A single stone pier needs a full-scale footing and a flat top.

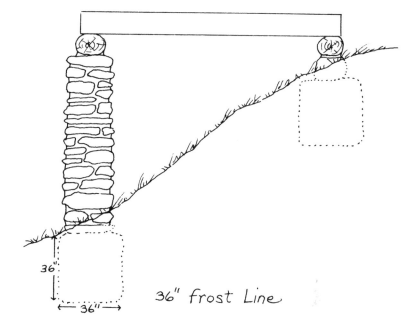

Fig. 7-3. A short pier needs the same footing as a tall one.

36"

36" frost Line

36"

If you use a short pier of smaller dimension than a taller pier downhill, the footing still needs to be as large as what you deemed necessary for the taller pier (FIG. 7-3). To some extent, you could vary the size of the footings to accommodate greater and lesser loads. Stone piers that carry the porch of a house (FIG. 7-1) would not need as broad a footing as the piers that actually carry the weight of the house. The octagonal log house seen in this chapter has a center post that carries weight from every rafter and floor joist down to a column that needs an extra-broad footing.

Some people make an excavation and pour it full of concrete. If you need to save money or if your access won't accommodate a cement truck, you don't have to use that much material to get a sound footing. You need to dig a hole to frost level or lower, and make a solid pad of some sort to spread the weight of the building over a broad area of firm subsoil. You don't have to build the pad all the way up to ground level, however. Instead, you can begin to build the narrower column, or pier, directly off that slab or big rock at the bottom of the hole (FIG. 7-4, LEFT). As the pier tapers from the footing dimension to the pillar dimension, empty space is left in the excavation. You can fill this space with gravel or dirt.

Another effective approach to pier footing is shown on the right side of FIG. 7-4. Dig a large enough hole to the frost line and then fill it with small rocks, gravel, and big, ugly stones. A few inches from grade, make a cap of concrete or an equivalent stone. If you use a stone to cap the dry footing, lay it in a bed of mud to give it a good seat. Then build the visible column on the footing cap.

Your column or pier of stone and mortar will need a footing even if it carries no weight other than its own. It is amazing how many leaning towers of stone or brick you see along a highway in front of businesses or long, private driveways. The useless structures are supposed to mark the entry with a certain regal elegance. Actually, the combination of cars and stone piers is lousy for all parties concerned: the driver, the vehicle, the owner of the pillar, and the mason.

The stone planter seen in Chapter 13, FIG. 13-19, was hit by cars three times just during the time I was building it. The owner of the fruit stand wanted them to accent his business along the highway. They weren't hit hard enough to hurt anyone; it was more like parking miscalculations, where 3,000 pound going 5 miles per hour bumped a thin wall of rubble stone. It is no fun to fix something like this, over and

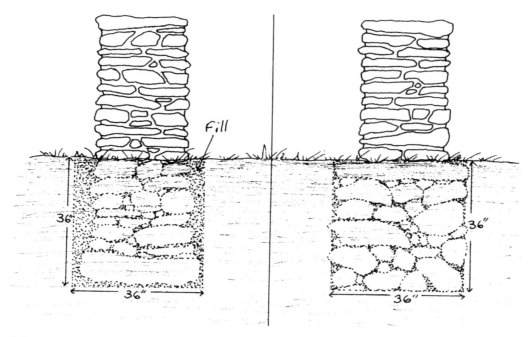

Fig. 7-4. *Two approaches to a rubble-stone footing for a pier.*

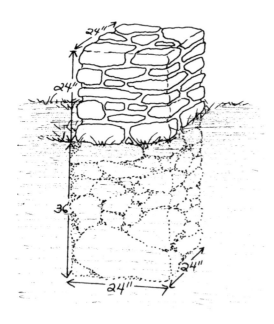

Fig. 7-5. *A short pillar that supports no weight doesn't need a footing broader than the pillar dimensions.*

over. It is actually not much fun to build it the first time if you know a car will soon be cracking the careful pointing job.

Even a 2-foot-high brick or stone pillar needs a footing. It hardly seems fair to have to build a 3-foot-deep footing to keep a 2-foot-high mortared pier from heaving and breaking, but that's what you would need to do if the ground froze that deep. That's another reason why these decorations of masonry are a poor idea: they take a lot of work for the amount of gratification. Even if you only stack 20 stones, you still need to go through all the motions, like driving to the cement place and getting your pants all dusty. A small job takes as long to prepare for as a big one. If you try to skip the footings on your stone driveway markers, you can add maintenance to the initial work load.

A short column supporting only itself wouldn't need a footing any wider than the pier dimensions, as shown in FIG. 7-5, which assumes a frost depth of 3 feet. With no frost, it would still be a good idea to dig down 1 foot. Here the soil will not compact as easily, and there is less chance of surface erosion undermining the pillar.

Laying Out
Footings for Stone Piers

The positioning of your stone pillars will be determined by the size and weight of the structure to be supported. Suppose you are going to build a workshop 10 feet wide and 20 feet long on level terrain. You will need six piers to give support every 10 feet. If you had two, exceptionally strong logs 20 feet long, you could possibly get by with only four piers—one at each corner of the rectangular building. The piers support the sill logs, which support the floor joists, which support the deck, which supports the walls, which support the roof structure. Good carpentry is like masonry: the components are stacked, with lapped joints, relying more on gravity than fasteners for integrity.

FIGURE 7-6 shows how you might lay out the foundation piers for a 10- x -20-foot shed. The two pillars at the top of the drawing are centered for the log beam that spans them. If you eventually wanted to complete the stonework between pillars, this positioning will get you a stone wall that juts out beyond the siding of the building. If you want all the stonework to come just short of the siding so the siding can lap down on the stone, then you should keep the pillar flush with the outside of the beam, (FIG, 7-6, BOTTOM

TWO CORNERS). In the illustration, the inner square at the corners is the aerial view of the pier; the outer square is the footing. The pillar should always be centered on the larger footing, if possible.

You need to know where everything is going to go before you can even dig the hole. If the building will have no siding and no other stone between pillars, you have a lot of leeway in the positioning of four piers. Logs can extend beyond the stone pillars and floor joists can overhang the beams. If you need more than four pillars to support your rectangular shed, the alignment is more critical. If you are building something fast and funky with crooked sill logs, you might need to lay out the piers with the crooked log lying on the ground in the right spot. Mark the excavation hole around the log, then push the log out of the way and dig.

When you build the piers, they should be in the right spot for the log. This is actually backward logic. The log will wear out before the stone, and someday your grandchildren will be looking for a crooked log to fit across the old stone foundation. On the other hand, if the stone piers only take a day to build and wood is scarce, there is no reason not to use the crooked wood.

Short rubble-stone piers that aren't supposed to be a showpiece can be built fast. I have done nine

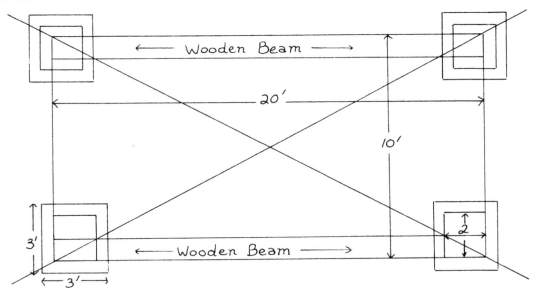

Fig. 7-6. Possible layout for a rectangular building on four piers.

in a day for a small house. This is one area of stonemasonry where you can use rubble, and it is the cheapest, fastest way to go. If your place is cluttered with stone, you could gather some and lay the piers in the time it takes to make the money to buy the blocks and to drive somewhere to get them. Piers in the center of larger buildings won't be seen, except in a crawl space, and they can be slopped together, as long as the result is stable slop.

If you are working with uniform lumber and building something that must look straight, you should get the piers and footings lined up so that the siding will come down and lap over the top edge of the stonework. To achieve this effect, you will need to use lines to lay out the piers.

If you know the long and short side dimensions of a rectangle—in this example 10 feet × 20 feet—you can calculate the diagonal distance with the Pythagorean theorem, which states: the sum of the squares of the two sides is equal to the square of the hypotenuse. In this example, (10 × 10) + (20 × 20) = the diagonal measurement times itself. To find the diagonal number, you must hit the square root button on the calculator after punching in the big number. In this example, $10^2 + 20^2 = 100 + 400 = 500$. The square root of 500 is the diagonal distance. The square root of 500 is roughly 22.36, which translates to 22 feet 4 5/16 inches. The procedure for turning hundredths of a foot into sixteenths of an inch is explained in Chapter 3.

After you know the diagonal number—in this case 22 feet, 4 5/16 inches—you can go to the job site with hammer and stakes and lay out the exact corners of the building. Using a helper and two long measuring tapes, place two stakes in the ground 20 feet apart where you want the front wall of the shed to be. Hook a tape measure to each stake so it starts at zero. Pull 10 feet from one stake and pull the diagonal distance from the other tape. Where these two distances intersect is the third corner. Reverse the ends of the two tapes and pull the same distances to find the fourth corner.

The tapes must be pretty level for any kind of accuracy, and the stakes need to be plumb down from the point where the tapes cross. If your site is cleared and fairly level, it doesn't take long to get these four stakes in the ground. The trouble is, you need to es-

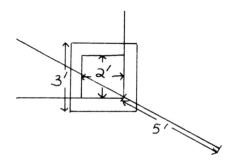

Fig. 7-7. Corner layout using expanded diagonals as a distance reference.

tablish another set of references so you can dig holes where the corners of the building will be.

A fast possibility is to pull a string between diagonals. Keeping the string straight, make a mark on the ground exactly 5 feet beyond each corner (FIG. 7-7). With a big hammer, put in a stake at each of those points and make sure it is straight up and down. Then you can pull the corner stakes out and dig the holes. It is helpful if the reference stakes are all at the same level. You can mark their tops level with the aid of a water level, as described in Chapter 3. This gives you a string to measure down from later when you are stacking the rocks. If you stop the first pier 7 inches from the string, for instance, you would build the other corner piers so they also stopped 7 inches from the string. This way, all the piers are level with each other, which is what you want for a foundation.

Once you know your corners, you also must know if the stonework is to stay within those corners. If it is, as in FIG. 7-6 BOTTOM, the footing should extend beyond the corner to keep the pier centered on the footing. The dimensions shown are typical: 24-inch walls on a 36-inch-wide footing. This would mean the footing would extend 6 inches beyond the corner. It isn't worth having more lines to describe that perfect 36-inch- × -36-inch footing. Measure 6 inches out and draw a new square on the ground and start digging.

FIGURE 7-8 shows an end view of the proposed 10-foot-wide shed. The pier on the left is the more stable arrangement, because the log is centered on the pier. The pier on the right is arranged so that siding might come down to cover the log and about an inch of stone. That wouldn't work on the left pier. If you have the stakes 5 feet out from your outside corner,

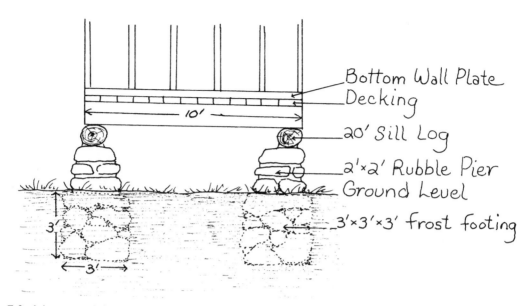

Labels on figure:
- Bottom Wall Plate
- Decking
- 20' Sill Log
- 2'×2' Rubble Pier
- Ground Level
- 3'×3'×3' Frost footing
- 10'
- 3'
- 3'

Fig. 7-8. A beam can't be centered on a stone pier if the siding is to lap on to the stone.

when you are ready to lay stones, measure in along strings between diagonals 5 feet and hang a plumb bob from the line at that point. The string must be very taut so the plumb bob does not cause it to sag. For the string to be very taut, the stakes must be very stable. With two plumb bobs hanging still directly above the true corner, close one eye, sight across the lines, and bring them into alignment. The plane described by that alignment is the outside face of the stones.

For short piers of crooked rubble, it is not worth a lot of fussing to get lines for all sides of each pier. If you know the corners, you can keep the job straight enough. You can lay the first course or two on a perfect square drawn on the footing, with a 2-foot framing square. If this simple building was to be larger and needed more than four piers, or if there was no 20-foot log, only 10 footers, you would need piers in the middle of the length. You could place them along a line around the corner piers. If you build your four corner piers first, you can use those piers to string lines that will locate the right position for the other pillars. The higher you need to go with a rubble stone column, the more you need to be accurate in your layout. On steep ground, you will have tall piers. The log octagon under construction in FIG. 7-9 has piers 1 foot tall in the back and 8 feet tall in the front. Of course, they come to the same level.

If you plan to add to your building in the future, you might want to build the piers so they leave plenty of surface protruding beyond the plane of the original walls. This surface will give you a place to rest beams that will carry floor joists for the addition, and you won't need to build a new pier there. You should have a good idea what you eventually want to do with the building before you dig the holes. If the building is to rest on stone columns forever, without additions or underpinning, then merely center your columns on your footing and center your beam on the column (FIG. 7-1). This is the easiest and perhaps sturdiest arrangement for getting a wooden building off the ground. In this approach, you can taper the columns to the top to save work. The top stone should be centered, and be large enough to give good bearing for the wooden beam (FIG. 7-10).

There is no particular reason to build square piers for foundations unless you intend to eventually connect them with more stonework. The 24-inch size often mentioned is something of a minimum for stability with rubble stone. There is no reason to be confined to that or any other dimension. If your stones don't have corners, let them stick out randomly, but make sure that the randomness contains within it a plumb section of stable stonework with a 2-foot diameter. You can taper toward the top if you don't start to do

Fig. 7-9. A log on stone piers on a steep hill.

it too soon and create a shaky situation. The tendency is to skimp on the size of a stone column if you feel there is no border line. The center of the top of your pier should be fairly centered on your footing.

Laying the Stone Pier

After you have footings for your stone piers that come up to ground level from below the frost line, you are ready to begin the stonework. If your footings are about 3 feet × 3 feet and in the right position (check with a plumb bob off the diagonal lines) mark a 24-inch square with chalk on each footing. It is possible to lay the pier without mortar if the stones are big and fit well or if the piers are very short—like one stone high. I have encountered many old buildings built on dry piers that are still in fine shape.

There is some danger of stones cracking from excess weight where there are hollows under stones. You can minimize this danger by using strong, flat stone, and filling any hollows with smaller stones. In some cases, a dry-laid pier is superior to a mortared one because the space between stones prevents ground moisture from flowing to the wooden building (called *wicking*). Even if just the top stone is laid dry, it will stop the capillarity of water from below. If you do dry-lay the piers, you should use a big stone at the top. You will have to plan the piers to allow the capstone to come out at the right height. The weight

of the building on the capstone will stabilize all the stones below.

FIGURE 7-11 shows the use of a homemade water level to get the tops to the right height. The level of the water in the jug or bucket is the same as the level in the plastic tubing, provided there are no bubbles in the tubing. Set the water reservoir on a temporary pillar or chair near the center of the work area so you won't need an extremely long length of tubing. Get the water level at the height you want all your pillars to be. As you get near the top of a pier, check with the water level and measure down to see how much higher you need to go.

Suppose your pier is 8 inches short of the water level. You would want to find a nice stone that is 7 inches thick so that, when it was laid in a mortar bed, it would come out right. If the water level tells you that the pier is 2 inches short of the top, you are in trouble. It is hard to find a strong capstone that is so thin. You would need to take some stones out and make a course that was 2 inches taller and skip the little stone on top.

With rubble stone, it is a good idea to have your pier caps picked out before you start, and custom-make each pier to end with that good stone. The sill log or beam should rest on a solid, flat surface that isn't likely to get moved by a bump. Too often, an unplanned pier will end with a small shim stone, and the

Fig. 7-10. *A pillar can be tapered to save materials.*

weight of the building can crack it and displace the pieces.

Using Mortar To Lay Stones

A stone column has so much surface area compared to its volume that it isn't worth laying the stone and pointing it in different mortar mixes. Instead, do the whole pier in a mortar mix for pointing. It will be stronger and more adhesive than the bedding mix illustrated in Chapter 6. Three parts sand to one part masonry cement will work fine.

If you have never worked with mortar before, begin in an inconspicuous place—like a pillar that is in

the back of the building, away from the access. Make the mortar as dry as you can so it will still form a ball without crumbling. Make sure the stones are clean and moist (not wet) before you begin to lay them in the first mortar bed on the footing. The mortar is mostly a temporarily liquid shim. Use it to give even support and a level bed for each stone. Don't be tempted to glue your stones with the stuff; instead use the mortar to seat the rocks in the position they naturally want to go. If you have had experience laying stones without mortar, you'll do much better with mortar. You will understand its role better and mostly use it to fill the spaces between the imperfectly fitting stones.

Mush the first stone down in the bed of mud, grab the mortar that squishes out past the face plane with a trowel, and put it where it is needed. After fitting a whole course of stones, fill the vertical spaces between them with mortar, and put a rough cap of mortar over the whole course. This procedure will allow the next course to begin on a flat, levelish place with a bed of mud about 1 inch thick. Start each course of the pillar with a large cornerstone, and react to that with other stones to get up to the same height. On the next course, start the large cornerstone in a different corner of that pier. This will keep joints overlapped naturally and distribute the strength and looks evenly. Use the best looking stones on the most visible sides of the piers.

If all your stones were 24 × 24 inches, you would merely stack them on top of each other and be done. Instead, you must try to create conglomerate rocks 24 × 24 inches, and stack them on top of each other in such a way as to cover your cleavages. The courses certainly don't need to be the same height. They will probably vary between 3 and 12 inches. The courses can taper inward, as long as all stones below have stones on them. Don't let small stones get stranded on the wall by a higher course that doesn't cover part of them. Every stone should have a stone on top of it, except the capstone, which should be pretty big and have the weight of a building holding it down.

As you stack your rocks in mortar, don't go so high that the weight starts crushing the mortar joints at the bottom. If you are building tall piers, you will only be able to stack a few feet in a day on each one, and switch to another one. It is good to be working on several at the same time so the mortar in all of

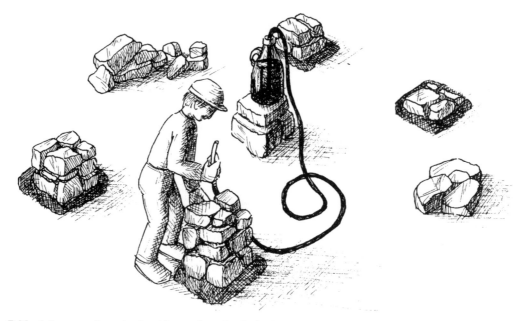

Fig. 7-11. A homemade water level is used to check the height of piers.

them gets a better chance to set up. If you dry-lay the pillars, you can go as tall as time allows in a day. As you lay the stones, cut off excess mortar from some joints and stuff it in other ones.

If you keep the mud fairly dry, it won't smear the stones much. If you use white masonry cement (white portland and lime, 1 to 1) it will stain them less. If you make the mix as weak as is practical, it will stain the least. Remember, if you can get the rocks to stand up at all, it is going to be beautiful. The only way you can mess up the looks is by letting the mortar get all over the stones, or making the mortar too much of the pier.

If you get a large space between stones within a pillar, you need to put small stones and mortar in the space. If you fill too large a cavity with mortar, it will start to slump out of the wall. Mortar needs something to grab onto other than itself. Concrete is mortar with something to grab, and it does much better in a large piece than mortar.

If you get too much pressure on the wall, you will see some of the water dripping down the pier. The water is getting squeezed out of the mortar, and that's not good because the mortar won't have enough water for proper setting. This is more of a problem when you use big wads of mortar to get a wall filled quickly.

At the end of the day or at the end of a batch of mortar, go back over the visible mortar joints with a thin trowel and pack them in good where there are little gaps. Use the tip of the trowel to scrape off the rough surface of the mortar and get it down to a smooth, uniform surface. Cut away chunks of mortar that have curled against a stone. Do not dress the joints if the mortar is at all tacky. It should crumble away when you plane it down. In some cases, it looks great to leave the rough mud protruding between stones. I'm of the school that thinks gray cement is the worst color ever invented, and I try to minimize the sight of it. I like the way rocks look when they fit together. Certain mortar colors are much less offensive than portland gray. Using white cement and yellow to brown sand gives a mortar that looks more like sand or dirt than cement.

If you want to be fancy and get a tight mortar joint, brush the spaces between stones briskly with a small paintbrush after you scrape them to a nice surface. A little experience will tell you the right time to dress the mortar. If it is too early, a brush will pick up a lump of cement as you go over the joints. If it is too late, do it anyway. It will be harder to scrape the crumbs away the next day. If your stonework is not meant to be a show piece, but rather a cheap, fast way to get a chicken coop off the ground, you could go over the pillar with a stiff broom to knock off the stray blobs of mortar and tighten up the joints. The brushing action brings a little extra lime and water to the surface, giving the pointing a slick crust.

BUILDING A COMPLETE STONE FOUNDATION

If your building is to have a continuous stone foundation instead of separate piers, you will also need a continuous footing. The advantage of a complete foundation, other than greater strength, is that you get an enclosed crawl space—or basement, if you make the walls of the foundation 8 feet tall. Even if your foundation is low (FIG. 7-12) and you don't get a usable crawl space, there are advantages in closing in the whole foundation. The stone wall in FIG 7-12 goes down to a footing, which extends below the frost line. Obviously, the wind can't blow under a house like this, and bugs and rodents would need to burrow down 3 feet to get under the wall.

A full stone foundation will give any building a more permanent feel and look. The extra labor is clearly not worth the effort for some buildings. A complete foundation wall might make the most sense when it is tall enough to create a basement. In this case, the foundation walls become the basement walls. If you are faced with the possibility of having a building 5 feet off the ground on a stone foundation, you might want to build on a few more feet and get a useful basement, instead of a semiworthless crawl space. Lots of old houses have basements with ceiling about 6 feet tall. People are often tempted to remove more of the dirt floor to gain headroom. This would undermine the footing and weaken the structure.

FIGURE 7-13 shows shortened end views of the same building with three different conditions in the basement. On the left is a wooden building on 3-foot-tall piers which sit on 3-foot-deep footings. The arrows show some of the forces. The building pushes straight down on the footings, the footings push down on the dirt, and that pushes dirt off to the sides if the sides

Fig. 7-12. A new wooden house on a stone foundation.

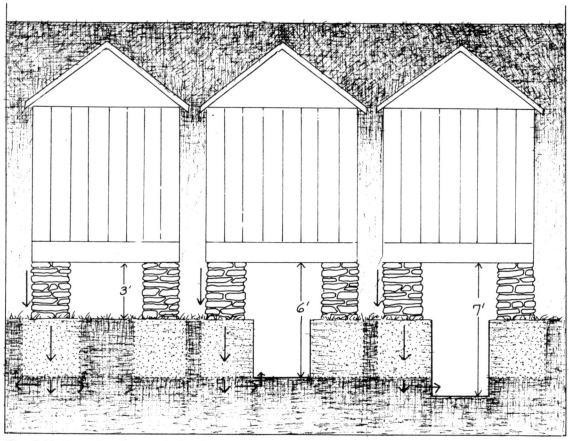

Fig. 7-13. Removing dirt to make headroom in a basement can undermine the footing.

are softer than the bottom of the excavation. There is such a great bulk of dirt to be moved out of the way before those footings can go anywhere that they don't.

In the center of FIG. 7-13, someone has figured out that they can gain valuable storage space by removing 3 feet of dirt. Now they can stand up under the old barn. The arrows show that the force and pressure on the subsoil now has less resistance because it has been removed. The dirt on the inside edge of the footing could get pushed up a little, and the footing could start to lean a little. It is the kind of thing you could probably get away with, but should never do.

On the right of the illustration is the next step in undermining a footing. A 6-foot-tall person got tired of banging his head on the ceiling in that new space, and decided to dig out another foot. The excavation has now gone beyond the depth of the footing, and as the arrows show, pressure could easily push the dirt out sideways, which would give the footing somewhere to move to, and that's not good for a building.

The worst extreme of this common blunder is when you start to remove the dirt on the outside of the wall too, until your stone wall and footing are sitting on a short wall of dirt. If you remove the dirt from the sides of a footing, then the footing becomes part of a wall with no footing. A pre-Civil War stone house in my area was recently leveled to the ground when someone tried to dig the basement deeper with a backhoe. The backhoe operator was saved by the frame of the machine, when hundreds of tons of stone came crashing around him. If you are worried about the stability of an old stone or brick house, add more dirt to the basement; don't take any away.

Layout for a Stone Foundation

A stone foundation that is not going to be a basement wall can be laid out the same way as a set of piers. You really only need the outside lines for the wall face. A foundation as seen in FIG. 7-12 will never be seen on the other side, so it can be pretty ragged, with no attention given to faces or pointing. If this house had a full basement, then the stone wall would be seen on the inside in the basement.

The footing should extend about 6 inches beyond the wall. The face of the wall should be plumb so that it can all fit flush with itself. The back side of an unseen wall could taper toward the top to save materials (FIG. 7-14), as long as there is no great structural requirement on the wall. If the tapering wall confuses you or makes the stonework slower, by all means make the wall a uniform thickness.

As you get more familiar with the medium, you will realize where you can leave out stones without jeopardizing the structural integrity. On a complete stone foundation for a relatively light building of wood, the wall and footing size could be scaled down to the load. On separate piers, there is much more weight to be spread out on the topsoil than with a continuous wall. For a wood frame house of one or two stories, the footing for a stone foundation could safely be 2 feet wide, as it would for a conventional block or concrete foundation. The wall on the footing could

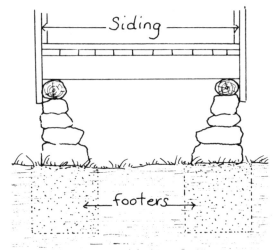

Fig. 7-14. Foundation walls could taper in on the inside where they aren't seen.

be squeezed to 18 inches successfully. This method would save a lot of material compared to building a 24-inch wall on a 36-inch footing. A rubble-stone wall is already so thick that footings do not need to be twice or three times as wide as for a narrower block wall. Never let the footing be narrower than the wall it carries. It should be wider, unless the wall is very thick and not carrying much weight beyond its own; in which case, the footing could be the same dimension as the wall.

The top of a stone foundation wall will not be seen because of the building on top. You don't need to take as much care in fitting the top, and you can fill depressions with small stones and mortar to bring things up to level. Keep the faces of the walls plumb by eyeballing across two plumb bobs, in line with the wall, or stretch a string from one end of the wall to the other and bring the stones in the middle to the string.

In this second method, build up the ends, or corners, of the building first, using a level. Then pull string around outsides of the corners, and you will have a meaningful line to guide you quickly through the rest of the course without a level. Don't let the line actually touch the masonry; it has to be held off 1/2 inch or the stones you set to it will start to push the line out and you will hardly notice until it is too late.

Rubble stone is not good for use in holding a line. You can use brick or block, but you need a square corner. It is best, then, to not use the stone wall itself as a reference because it is not accurate. I like to have a plumb bob at each corner to build the corner itself, and have pairs of parallel lines to sight across to check the positioning of face stones. This much can be had from the expanded diagonals described earlier. Four stakes, two strings, and two to four plumb bobs are all you need. It is nice to have the strings set to the top level of the proposed foundation wall. Braided nylon mason's twine is amazingly strong stuff. Don't be afraid to pull it tight, especially when checking the top level of a wall. A loose string will sag, especially over a long distance. If you follow the string with your stones, the wall will also have a sag in the middle.

Another layout possibility for foundation walls is to set eight stakes in the ground, in line with the walls, but beyond the corners (FIG. 7-15). These stakes can be established by two people once you have located the true corners. The lines they carry can be used as

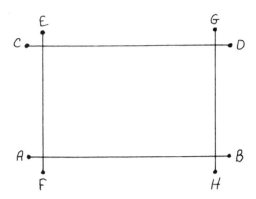

Fig. 7-15. A layout using eight stakes shows four corners.

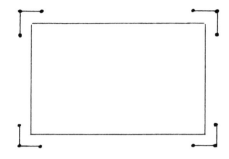

Fig. 7-16. A layout with 12 main stakes and batter boards can show all needed lines of a rectangular building.

sight lines for each wall, in conjunction with lines chalked on the footing plumbed down from the strings. These eight stakes will only tell you where the outside faces of your wall need to be. To locate inside faces, measure off the front face the desired thickness.

If you want more accuracy on the inside face of the foundation wall—if it is to be part of the visible interior wall—the layout in FIG. 7-16 could be the best. Here there are three stakes beyond each corner, connected with batter boards. Make marks at the proper place along the boards, and string lines to show various planes in the construction. The boards are long enough to carry lines for the inner edge of the footing, the inner wall face, the outer wall face, and the outer edge of the footing. If you need all that information, complete batter boards are worth the effort.The arrangement in the illustration could require 52 separate pieces of wood, including the main stakes, the horizontal boards, the braces, and the small stakes to attach the braces. If your walls are low and not very long, and your building site is level, you might not need to brace your stakes. If the wall is tall and you want your stakes to carry strings to tell you where the top of the wall is, the stakes will get pulled over by a tight line and would need good bracing.

It is quite possible to build a stone foundation as part of an otherwise wooden wall. This implies a floor at grade, probably on a concrete slab. If you decide to use a partial stone wall in your building, the inside face of that wall will need to be uniform. If you want the inside and outside faces to show, there is no way

to insulate the wall. Forget about splitting a rubble wall in two with Styrofoam in the middle. You could do that with thick brick walls, but not with rubble stone, unless you want walls more than 3 feet thick. If you don't care about the insulation problem and you want both sides to show, you will have a slow job on your hands. It is much harder to fit rubble stones when both sides of the wall will be exposed.

If you decide to insulate the stone wall, you could build it against a plywood form or temporary wall. Set the plywood backing plumb with braces (FIG. 7-17). A wall built against a backing like this can be very fast. Establish the front face by measuring off the backing. The back side of the wall comes out smooth and flush when the forms are removed, giving a good surface to attach insulation to. A low wall built this way could be as thin as 12 inches, if the stones are good and many of them are 12 inches deep. I would recommend 18 inches of thickness for this type of building. You must determine the minimum thickness according to your stones. Naturally, if your stones were shaped exactly like cement blocks, you could build a wall as thin as a block wall. Don't build a stone wall so thin that if the pointing mortar falls out, the wall will fall apart. The longevity of stonemasonry can be attributed to its basic structural overkill.

If you decide to build your foundation wall against plywood braces or half-forms, it is helpful to have a fairly level footing to set the form on. If there are 3 inches or so of footing beyond the wall, you can set the form on the footing (FIG. 7-17). You can set something heavy, like blocks, against the back side of the plywood backing to keep it from kicking out when stone is set against it. Stones should not lean against the wood, but merely come to it, or slightly short of

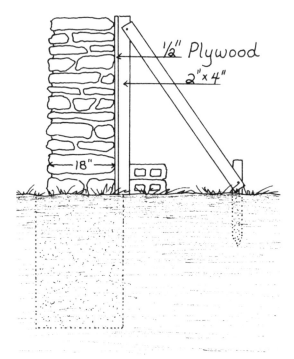

1/2" Plywood

2" x 4"

18"

Fig. 7-17. *Building a foundation wall against a temporary backing.*

it. Pack the remaining space with mortar. If the plywood extends above the intended top of the wall, you can snap a chalk line on the plywood to show where the wall should stop.

If you plan to lay stones to a half-form, you will be measuring from the inside out. It would be helpful to have a line on your footing that shows where to set the base of the plywood.

Suppose you wanted a building 20 feet by 30 feet, two stories high, with the first floor having a wall that was stone halfway up, and then wood. If the walls are to be 18 inches thick and built against temporary plywood backing, then the inside dimension of the walls would be 36 inches less in each direction, or 17 feet × 27 feet. To find those inner corners, find the diagonal of the inner rectangle. $C^2 = 17^2 + 27^2$. $C^2 = 1018$. $C = 31.9$ feet, or 31 feet 10 7/8 inches. Lay out this triangle on your site as explained earlier in this chapter and in Chapter 3. Get some references beyond the site so you can dig a footing and locate the corners again.

When you are sure that you have a rectangle that exactly describes the inside dimension of the wall,

make corresponding chalk lines on the footing and set your plywood forms to that line. If the footing is too rough to accept a chalk line, you can use a string pulled tightly along the footing to set the forms to. FIGURE 7-18 shows where you might set stakes for this layout, coming beyond the corners an arbitrary 5 feet to keep out of the way of the excavation. If you leave the reference stakes in the ground, you quickly can check the positioning of the plywood if you think it has been bumped out of alignment.

The procedures using measuring tapes, strings, and stakes might sound complex, and they are, compared to the rest of the job of construction, but it is time well spent and easy on the body compared to laying stones in the wrong place.

Openings in a Stone Wall

If you want a window or vent in your foundation wall of stone, it is best to have a wooden frame of the correct size ready to set on the wall when it gets to the height of the bottom of the opening. The frame should have the same depth as the wall, that is, 18 inches. This depth is usually achieved by joining narrower boards.

When you set the braced box opening on the stone wall, make sure it is plumb and level. Bring a few large stones with square ends up against the form on either side to hold it in place. Don't lean stones against the window frame; merely bring them to it without pressure. You can anchor the frame into the mortar a few times with big nails. When the sides of the opening are stoned in, you must span the top with an arch, a steel lintel, or a long stone.

Some people have bolts waiting in their framework to attach to the masonry. I don't like to fit stone to bolts. I'd rather put fasteners where they can fit after the stone is in. A few cut nails into an obvious mortar joint through the opening framework will keep the frame from moving in or out. The stonework and the frame itself will keep it from moving any other way.

If you are building something for keeps, don't get into an arrangement where the wooden parts of a stone building can't be replaced. Especially don't allow the stone to depend on wood for its support. Even pressure-treated wood has a flash-in-the-pan lifespan compared to stonemasonry, so don't mix them structurally.

111

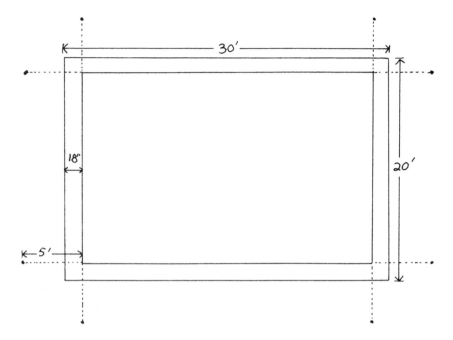

Fig. 7-18. *An eight-stake layout for the lines to set the plywood backing to.*

UNDERPINNING A STONE FOUNDATION

Often, a wooden building set on stone piers is closed in underneath with stone as an afterthought. It is nice to have a continuous wall under a house, but making a full stone foundation can be tedious compared to building on piers. If you can anticipate the desire to close in the foundation eventually, you can save yourself some hassle by building a complete footing from the beginning. Generally, these days footing work is done with machinery. It's not much extra work for a backhoe to dig the whole footing than to just do the pier footings. If you make a complete footing, you can proceed to build the piers, followed by beams and a deck.

This method will enable you to get the structure up quickly. Years later, when your chores are all done, you can start to lay a thin, barely structural stone wall between piers. Because the underpinning stone is sitting on a nonheaving footing and not carrying any weight, it can be as thin as what you are able to stack, with no regard for the looks of the back side of the stone.

Underpinning is fast, gratifying work. It will keep the wind from blowing under the building and give you something that is very close to a real stone founda-tion, with much less work. If you can anticipate doing this, it helps to lay the piers with staggered courses, which will enable you to tie into the piers with the underpinning later, and not have an obvious seam (FIG. 7-19).

Unfortunately, most times you see underpinning, there aren't staggered ends on the piers or a complete footing. There is still much that can be done to add charm and integrity to a stone pier foundation or a foundation of block pilings. For a building on block pilings, you can place braces next to the block pier, remove the blocks and replace them with stone. If nothing else, you can build a dry wall right on the dirt to come up to your building. Even with lots of cracks, a dry wall under a trailer edge will keep the air from moving quickly under the building.

FIGURE 7-20 shows the house that was built on the foundation seen in FIG. 7-9. The owner-builders, after living in the house for years, decided they wanted more space. The easiest space available to them was under the house. The roof was already on it, and there was enough room to stand in the front half of the space under the house. One of the residents decided to dig in the basement area to make footings so that the stone columns could be connected with more stone. The unseen center pillar of the log octagon needed to be completely undermined to give any semblance of a basement floor. This was cleverly done one-half at a time and replaced with a massive concrete base and a wooden post. After shoveling ton upon ton of dirt from under his house, the owner decided he'd had enough heavy labor, so he hired me to fill in between his stone piers.

The pillars were backed up with a temporary wood wall (FIG. 7-21), which made it easy to keep a

Fig. 7-19. Piers with staggered sides can be tied into nicely later.

Fig. 7-20. Closing in between columns will give this house useful basement space and retard heat loss.

flush back on the wall. The wall was later to be insulated on the inside. Notice in the same picture the center opening had been made square. This represents a change of mind. The builder was going to fill in the space with more logs, and so filled in around the piers with stone to get straight ends. For me, this made a challenging place to try to blend the new stonework.

Since I was working alone, I opted to use a wheelbarrow. Notice the section of octagon on the left. The window frame was already in place, suspended by nails in the bottom log. The frame made it much easier for me to lay the stone than if I was told to leave an opening a certain size for a window.

Because this underpinning was over 6 feet high in places, the idea of building a thin wall was out of the question. Instead, the new wall was kept to about 18 inches. The wooden wall behind the masonry was not attached to it; it was mostly a mortar dam and a guide. Because the boards were to be removed later, the stones could not (and did not) depend on the wood for support. Notice how the course is level, front to back and side to side.

The mortar was left rough and recessed on this job for several reasons. One, the owner didn't have enough money to pay for a longer job with finish pointing. Second, the existing piers were so rough in finish that there would be no point in making the new work look alien. Third, this unpointed look has aesthetic appeal for many, as it is reminiscent of dry-laid work.

FIGURE 7-21 shows a section of the octagon ready to be stoned in. Notice on the right side, the stone work had been left in a straight line. I removed a few stones to break up that unbroken vertical joint. The other side was rough enough to tie into inconspicuously. Unfortunately, the top two stones on the right were too big and structurally important to knock out of the wall.

After the footing was moist and dirt free, I laid the bottom mortar bed in and started the bottom row of stones. Because the ends of this little wall were already in place (the old stone columns) no particular order of laying was imposed, except to tie in (overlap) to the sides, and that meant paying attention to the height of the existing rocks. I laid the front face first, to the height of the existing stone to its side. This method allowed the stone above to sit on both the old

and new walls, and this is what I mean by *tying in*. I filled the space behind the face to the wood brace with small stuff laid flat with mortar between, until it reached the height of the face. FIGURE 7-22 shows several aspects of that procedure. To the right, the course is finished. In the middle, there is still a gap behind the face stones. It took mortar, followed by another layer of small stones, then more mortar to get the wall flat and level the whole way across. I had to lay the little stones in the back without creating any wedge forces on the temporary backing.

Because I knew this wall wouldn't have a finished mortar joint, I saved money using concrete sand instead of mason's sand. The concrete sand is coarse, with some pieces larger than what a bricklayer would want. The rough sand has more variety in grain size, which means it needs less cement for strength. We mixed the cheaper sand with four parts masonry cement. This mortar is a bit clunkier on the trowel and less sticky than regular type N mortar, but quite adequate for rubble work of this sort. Stones don't get buttered in rubble work; they get bedded.

Here's a tip for masons, as well as handymen, for wintertime building: Concrete yards will sell you concrete sand that is steam-heated. If you buy a ton of the hot stuff, you can get through a masonry job during slightly freezing weather. Frozen sand can be a hassle to work with.

FIGURE 7-23 shows that section of foundation wall progressing up the side of the window frame. Notice that the stones abutting the frame are square-ended for a close fit. I used small stones in this area for the most possible tie-in with the existing stonework. The small face stones are able to poke right in the crevices between stones on the old work. Bigger stones used here would never have fit in those indentations, and the result would be uncrossed joints and big gaps. Notice also that the stones under the windowsill are rather large. Where the wood starts is like the top of the wall and deserves the extra visual and structural elements of a big stone. The stone wall is most vulnerable under a windowsill because there is no weight on it. The weight of the house is carried over that area.

FIGURE 7-24 shows that section of wall complete. Compare this photograph to FIG. 7-21. Notice how the straight line on the right of FIG. 7-21 has been obscured in the finished wall. Notice that the window

Fig. 7-21. It helps if stonework can be done around the opening frames.

Fig. 7-22. The space behind the face stones is filled in pieces.

Fig. 7-23. New stones tying in with old. Square ends are brought against the window frame.

cannot move and doesn't need big bolts in the mortar anymore than a window in a frame house needs to be bolted. There are a few nails protruding into the mortar to keep the window from moving in or out. Treated wood was used for the opening frames in this stonework. This is a good idea since a foundation wall is prone to wicking moisture, which can soak into any adjacent woodwork.

Although the new stonework is very different from the old, it still blends with it. The rocks are all from the immediate vicinity. The initial stonework was laid in soil cement by a variety of people. There usually wasn't much chance of matching the style.

FIGURE 7-25 shows three sides of the eight-sided foundation closed in. The original stone piers now look like buttresses in the foundation walls. These piers were not planned as part of a complete enclosure. Even if they had been, the angle at the corners would be hard to do crisply with rubble stones. This much work took about four days with two people. Having the boards as backing already in place made things faster. The footings were also done beforehand, as was the stone gathering.

Flashing metal had been placed between the top of the stone wall and the bottom of the first log to block capillarity of water from the ground. It was cut back from view when the wall was done. Don't think you can lay a chunk of wood in mortar and maintain a complete seal. The wood will shrink some and the mortar will not be in contact with the wood for long. If you want mortar to make a tight seal with wood, you will need to attach a metal lath of some sort (hardware cloth or chicken wire will do) to the wood, and then attach the mortar to the metal. The mortar will hold on to the metal wonderfully, but it won't bond with wood.

For the next section of this project, the footing is clean, the wood frame for the opening is in place, and a pile of rocks is on hand. I laid the first trial course dry under the window (FIG. 7-26). The course is the height of the stone adjacent, which was already there.

This kind of work doesn't require lines or levels. It is usually a case of fitting things with what is already there, and making compromises between getting flush with a windowsill or keeping in line with old stones. What is there tells you where to put the stones. I pulled out the dry-laid trial course and put down a bed of mortar. I then put back the stones in the same position as for the trial course, and pulled mortar up their back sides with a trowel to jam the spaces full. Then I filled up the space behind the face with stones and mortar to that height.

When only one stone remained to be fitted under the window frame, I folded a piece of aluminum foil to fit the vacancy. You can make a template like this, give it to your helper, and let him walk through the stone pickings and bring back a stone that matches that size as closely as possible. This technique can save wasted trips with stones in hand. It is most useful for totally closed in areas that are oddly shaped.

To be able to do rubble masonry quickly, you must be familiar with the basic logic of the sequence and also be able to take mental photographs of a space. You can carry these mental shots to the rock pile, reach in the heap and say "There it is!", and come back with a stone that fills the space. This is the fundamental difference between this type of masonry and the kind with prefabricated units.

FIGURE 7-27 jumps to the last bit of stone to be laid in that section. The stone just under the trowel that is stuck in the wall is long, as was needed to cross a series of vertical joints below. However, the right side of that stone does not abut the window frame with a straight line. You can see that this stone created two small spaces that can hardly be filled with anything. It was a compromise move. Notice that the top course was brought level with some small stones. It would be good at this point to try to find the biggest stones that would still fit in the wall and fill the space with the fewest pieces.

When you get to the top of a wall that comes up to a ceiling of any sort, you must trial-fit the top stones, taking note of the amount of space behind them. Then, pull out the top few stones, jam in mortar and small stones in the back space, and push the face stone back in. This is the reverse of the usual procedure of filling the back after the face, but there is no way to fill the empty space behind a face at the top once the top stone is in, so the back must be filled first or a cavity will remain. Make sure some mortar squishes out the face to be sure the space is filled.

I was lucky to get three fairly large stones in to

116

C-1. These massive stone columns support a floor beam as well as add rustic charm to a country house. (below) The same columns, as seen from their opposite sides. These columns moderate the temperature in a passive solar greenhouse *(Will Lane)*.

C-2. (above left) The uncut mountain sandstones in this chimney are fit tightly so only a small amount of black mortar shows.

(above right) From the inside view, the backing for this soapstone stove suggests a stone chimney with considerably less work.

C-3. (left) This sandstone chimney and hearth is kept two inches from the wooden wall so that air can circulate behind it *(Will Lane).*

C-4. (above) This well-built dry wall holds earth against a stone garage wall.

(right) The same stone retaining wall seen from the end. The top stone is shown being placed onto the wall, which was built in one day.

C-5. A curving, dry-laid retaining wall of rubble stone holds earth against a solar house.

C-6. This colorful stone wall in white mortar has rocks of all types and sizes *(Will Lane)*.

C-7. (above) These limestone columns, salvaged from an old chimney, add support and thermal mass to a solar timber-framed house.

(right) After a wall has been built and pointed, a bristle brush and water will get a job this clean if done at the right time.

C-8. (above left) A ceramic foot emerges from the mortar of the landing step.

C-9. (above right) This one-room stone house was built the old way by two men in six months (D. Kennedy).

C-10. (right) This dry-laid wall accents a driveway. A sculptor chiseled the address into a protected stone.

C-11. (right) The reverse side of this fireplace is a Trombe wall in a greenhouse. The circles in the lower left and upper right corners are air vents *(Alan Paulson)*.

(below from left to right) The upper left corner sports carved brownstone from an old building. A relief arch over the lintel stone helps carry the tall chimney. A brass horse has its back legs and tail firmly attached to this artistic masonry.

C-12. An extremely tightly fit wall of rubble stone in black mortar.

Fig. 7-24. The section is closed in with stone and a window.

Fig. 7-25. A house gets a stone foundation with a dry-laid look.

fill the last spaces (FIG. 7-28). The squarish one at the top was the kind of find that makes a rubble stacker slap his dusty pants. It fits fairly square against the wood window; it crosses two joints below; it leaves an easy-to-fit left end; and it comes to the top of the wall, except for the little notch in the top log, which was filled with two small, flat stones. Stones that meet wood horizontally should have surfaces that slope away from the wall. It is easy to accidentally put a chunk of rubble at the top of a wall like this that will

117

Fig. 7-26. A trial run of face stones is put in dry.

Fig. 7-27. The inside must be filled before the face at the top.

Fig. 7-28. The remaining space is filled with the biggest stones that fit.

hold a puddle of water against the wood. Pay attention to the particulars.

Vigorously brush off the crumbs of mortar that remain on a wall the morning after you use the mortar. A hose and wire brush are a good combination for quick cleaning. Make sure the mortar is set enough to handle the water spray, but don't wait to clean up until it is as hard as rock.

STONE FACING

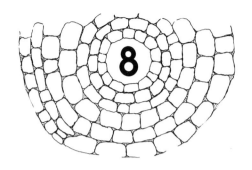

FACADES of stone have become increasingly popular over the past few decades. As people realize that a real stone wall is beyond their finances or physical endurance, they are turning to illusions of fine crafts. Stone facades are analogous to wood veneers: you get to see a lot more fancy wood if it is only 1/32 inch thick, and you get to see a lot more surface of stone for your masonry dollar if it is mostly a wall of something else with a stone covering. For me, it is impossible to fake a wall of stone with less than about 8 inches of depth.

LAYING STONE AGAINST BLOCK

An 8-inch-thick stone facade does not weaken an 8-inch-thick block wall, although it could not stand on its own. With thinner facades, the stone must lean on or be glued to the wall, which I think makes the block wall somewhat less stable. There is no exact cutoff point for when a wall is a wall and when it is a facade. Basically, if a wall won't stand on its own, it is a facing. Some 8-inch facings would probably stand alone, but would be easy to push over. If the stones happen to be shaped just like cement blocks, naturally they would make as stable a wall as the block.

The reason rubble walls are so thick is to compensate for the ill-fitting units. When you behold a legitimate stone wall, the stones you see are mostly much shallower than the wall. Perhaps they average 8 inches in depth. The same look, or effect, can be had with stone against block. You build what would be the front face of a real wall. The main difference is that you don't get to use the occasional tie stones you would use to tie together a stone wall. For this reason, you cannot assume that a block wall with a stone facing will have the same strength or integrity

of an all-stone wall or an all-block wall of the same thickness.

Brickties, or little strips of sheet metal with a rippled surface, are often left protruding from a block wall that is to have a stone facing. This method allows a physical connection between the facing and the wall. I would hate to feel like I was actually depending on these things to keep a wall intact. There is hardly a body of historic information about stone facing jobs. I feel certain that an 8-inch-thick stone facing, using a fairly adhesive (Type N) mortar on a good footing, would need no metal ties to a sturdy, porous block wall. The ties certainly won't hurt, unless you use them to do poor masonry. If the block wall you want to face was built without the protruding metal connectors, you can still do the facing job. As you get thinner than 8 inches, which is heading away from masonry, you will become more reliant on fasteners and adhesives. I'm not advocating using no metal ties; I'm advocating thick and realistic-looking stone facings that probably don't require ties to the wall they cover.

Having 8 inches of depth will allow you to stack flat stones on top of each other, like nature intended. The stacked facing will naturally look like a stacked wall. The other approach to facing walls is to tip very narrow stones on edge, but this never looks real because it is not built the same. Chapter 11 discusses the vertical use of flagstone in more depth.

Footings for Facings

All stone facings, thick or thin, need to rest on a footing that in turn sits on firm ground below frost level. If you know in advance that you are eventually going to face your block or concrete wall, it is smart to make an extra wide footing to accommodate the facing. Ideally, there should be 6 inches of footing beyond the edge of the facing. If your block wall is 8 inches and your facing is 8 inches, 12 more inches of footing would imply a footing 28 inches wide. Most modern residential construction is done with a 24-inch backhoe shovel, and I think the 16-inch-thick wall and facing could sit on a 24-inch-wide footing. This would leave 4 inches of footing beyond the wall on both sides.

If the block or concrete wall is load-bearing—that is, if it carries the weight of more than itself—it should be more centered on the footing, while the facing would come closer to the edge of the footing. Sometimes a concrete slab is poured over the edge of the footing and against a block wall. You could safely put facing on the concrete against the wall, even if the footing didn't actually extend beyond the facing. The facing won't carry any weight except its own if it is done after the structure is built.

FIGURE 8-1 shows three examples of a block wall on a footing, with a slab that laps on to the footing and a stone facing above the slab. The top illustration shows the safest approach, where an extra-wide footing is used to handle the facing. The middle shows the block wall is centered on the footing, and the rock facing comes to the edge of the footing. At the bottom, the facing is partially dependent on the tensile strength of the slab, because the footing has run out. This is plainly not as stout as the other two possibilities, but I suspect the slab alone would be adequate for the job, as long as it wasn't going to move. If the stone facing is 8 feet high, it will exert about 8 psi on the concrete. If the slab can't handle that, you'd better dig footings for your piano.

The illustration implies facing on the inside of the block wall. Facings on the outside of a wall would be more demanding as far as footing is concerned. You wouldn't want the facing to overhang the footing at all. There would be no concrete slab to back up the footing on the outside. Also, erosion is more of a problem on the outside of the wall. If the footing is marginal and the gutters leak a little waterfall onto the ground in front of it, the footing could crack, which would cause cracks in the facing. A facing job is much less forgiving of motion and stress than a 2-foot wall. If it moves, it is going to break.

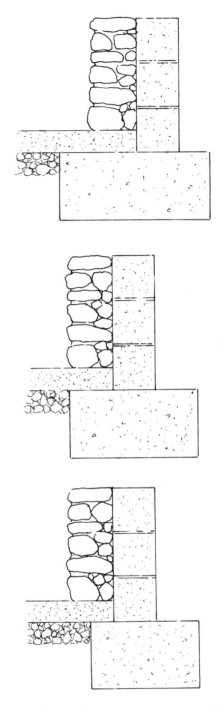

Fig. 8-1. Three different footing situations on an inside facing job with a slab floor.

Let's Face It

Facing a good interior block wall with stone is the easiest of stonemasonry jobs that require mortar. If the wall is plumb and flush, you will need no lines or levels, not even a tape measure. All that is required to keep the stones in the right place is a reference stick that is notched to show the proper depth of the wall. Abut the stick against the block, and set the stone to the notch on the stick. This method will keep all the stone faces on the same plane.

If the block wall is not plumb, you wouldn't want the facing to follow that crookedness. In that case, you would need some strings as guides to set the stone faces. The thickness of a facing on a crooked wall would be determined at the most bulged out part of the wall. That way, the facing would all be at least as thick as deemed necessary, and some of it would be thicker to even out the irregularities. By hanging a pair of plumb bobs near either end of the wall and about 1 foot out from the surface, you can establish a sight plane that is plumb. By aligning the strings with one eye, you can see where the wall bows in and out. If the facing job is small, you can quickly determine with a level where the high spot is on the wall.

Facing an inside wall is a cushy job, and a great place for a beginner to try stonework. There is no weather problem. You can work at night if you like. Nothing will get too hot, too cold, too wet, or too dry. Dust is the biggest hassle on an inside job. You will need to mix the mortar outside and minimize the amount of mortar blobs you drop on the floor. Cement dust is awful to breathe. The particles are so tiny, they get in the tiniest passages of your lungs and mucous membranes. Merely dipping a shovel into a bag of cement will raise a little cloud of the dust. It is so penetrating, it will go right through your clothing. Even if you are fastidious on a masonry job, you will be amazed at the sheer quantity of dust a pair of pants or gloves will gather. Possibly the ideal place to do an interior facing job is in the basement of a house. Chances are there will be no openings in the wall you choose to face. This will not only make the job easier, but it will allow it to look more like a real stone wall. Openings, or visible ends of walls, are the giveaway places in the illusion of a stone facade. At a doorway, for instance, you could see at the sides of the opening that the stonework was a facade. Inside

corners do not reveal the depth of the wall as outside corners do.

FIGURE 8-2 shows an 8-inch-deep facing, used to back up a wood stove. The inside corner does not reveal the depth of the masonry. On this facing, some stones are standing, but most of them are laid on their broadest side. I built the ends up before the middle, just like for a real wall. I built the facing in one week with 4 1/2 tons of materials. To save the customer money, I laid the stones in dark mortar, but I let the stones come in contact with each other. I kept the dark-colored mortar deeply recessed. The result is a very tightly fitted wall with a dry-laid look. This made a pointing job unnecessary, and therefore speeded the overall job.

An unpointed finish, which is safe on an inside job, greatly increases the surface area of the wall, so

Fig. 8-2. Rubble stone facing backs up a wood stove.

it is able to radiate its gathered heat more quickly. Pointing a wall like this would take nearly 50 percent more time without improving the looks of the facing. FIGURE 8-3 shows a close up of the technique. You can see the mortar if you look, but it is very low-key. Without rocks that fit tightly, including small stones to fill spaces in the face, you can't hide the mortar. These stones appear to be the roughest of rubble, yet I chose them carefully, avoiding those with outsloping sides and tops, and choosing those with a broad, flat base for stacking.

FIGURE 8-2 shows two other points of interest. One is the stove itself, which is a very efficient catalytic stove with viewing options. It is quite possible that catalytic combustion or an equally clean burn will soon be mandatory on all wood burners.

The other point of interest is the hearth the stove is sitting on. This was an experiment in black concrete. I mixed a normal batch of concrete for the hearth with added black pigment. Before the black slab was thoroughly set, I made shallow grooves in the surface with a thin trowel. When the slab got a bit stiffer, I pointed the grooves with a whiter mortar mix. The illusion is one of gray slate flagstone. The fake hearth required about one hour of labor and $5.00 of materials. A working wood stove manages to keep a fine dust on a hearth no matter what it is made of, so it is a disappointing place to spend a lot of time and fuss with stones. Experiments in concrete have a way of staying around to haunt you, but in this case, both builder and owner were pleased.

If the wall you are working on has a visible end or an opening, I recommend that you close in that space with finished wood. Then, at the ends, you would see wood instead of a stone facing against a block wall.

To create a convincing facade, a window opening in a block wall should have a wood jamb that extends beyond the block enough to cover the sides of the stonework. You can attach a trim board to the flush end of the jamb wood. The trim board will cover the gap between wood and stone, and cover an inch or two of the face stones around the window. The trim board will enable you to use a stone that doesn't have perfectly square ends to abut against the window frame. The trim will cover the little spaces. You also can use the end of the wall cover boards as sight lines or as string guides. If these cover boards are set plumb and square, the facing will follow. FIGURE 8-4 illustrates these points.

Facing for Added Strength and Mass

A facing job I recently did was 12 inches thick. It was in an underground house where more thermal mass was desired and where a wall needed to be stronger. In this case, the facing job could be considered structural. Eight inches of block and 12 inches of stone makes 20 inches of masonry. The other reason for the facing was to cover an ugly concrete wall with something interesting.

Fig. 8-3. Close up of the dry-laid look in a mortared wall.

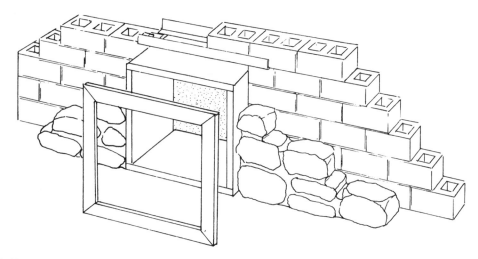

Fig. 8-4. *The sides of openings in a facing job should be hidden.*

The wall, which went to the ceiling, was 8 1/2 feet high × 17 1/3 feet long, or 150 square feet. The stone face was 1 foot thick so there was 150 cubic feet of masonry. At 150 pounds per cubic foot, the facing weighed 11 1/4 tons. It makes sense to calculate the quantity of masonry you will be using.

My 11 1/4 tons broke down as follows:

☐ Seven tons of face stones, of which 2 tons were very squarish for the ends of the wall
☐ 1 ton of fill stones, including shims and small face stones
☐ 3 1/4 tons of mortar.

A closer look at the 1 ton of fill stones, which is an essential part of the wall, would show about 2/3 ton of fist-size and smaller rocks of nearly any shape, which are used strictly behind the face stones to fill space. Broken bricks or block are great for this purpose if you can get them for free. The remaining 1/3 ton is about half little face stones, meant for the too-large mortar joints between rocks. These little face fillers should have some depth to get a strong hold in the wall. They are usually no larger than a pack of cigarettes.

The remaining 350 pounds breaks down as follows:

☐ 250 pounds of big shims
☐ 100 pounds of little shims

The shims are used to lift the back of a stone or to maintain a space at the face for pointing mortar later. It is darn hard to lay a rubble wall without this final 100 pounds (in this case) of little shims. On this job, two buckets was more than I needed. If you don't have a shim bucket on your job site, you will be digging around in the dirt looking for some while the mortar is setting.

FIGURE 8-5 shows the breakdown of components on this particular job. This doesn't represent any rules; it just happens to be what was used when I decided to keep track. Far from a recipe, this illustration analyzes the mixture in a fairly unusual wall.

The technique I used is explained in detail in *Fine Homebuilding* Magazine #27, in an article entitled "The Structural Stone Wall." I will explain briefly here.

Laying Stones in Concrete. Because it makes sense to use separate procedures to lay stones and point them, it is quite possible to use a very different mortar for the two steps. I found that I could stretch a regular 1:3 mortar mix to double its size by adding gravel. As long as the stones get coated with the mortar mix, the weak concrete will set.

Admittedly, adding gravel to mortar makes for a very clumsy mortar. The concrete is used to fill big spaces behind and between face stones. It is kept recessed enough so that a joint of regular mortar can be added at the surface.

The gravelly mortar has advantages other than lower cost. It tends to be less sloppy and more stiff

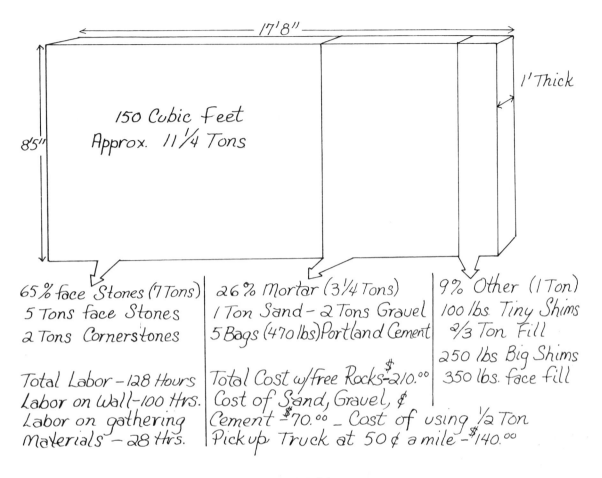

17'8"

8'5"

150 Cubic Feet
Approx. 11 1/4 Tons

1' Thick

65% face Stones (7 Tons)
5 Tons face Stones
2 Tons Cornerstones

Total Labor - 128 Hours
Labor on Wall - 100 Hrs.
Labor on gathering
Materials - 28 Hrs.

26% Mortar (3 1/4 Tons)
1 Ton Sand - 2 Tons Gravel
5 Bags (470 lbs)Portland Cement

Total Cost w/free Rocks-$210.00
Cost of Sand, Gravel, &
Cement -$70.00 - Cost of using 1/2 Ton
Pickup Truck at 50¢ a mile -$140.00

9% Other (1 Ton)
100 lbs. Tiny Shims
2/3 Ton Fill
250 lbs Big Shims
350 lbs. face fill

Fig. 8-5. The breakdown of ingredients on a facing job I did.

than mortar. You can stack much higher before the mortar gets squished by the weight above. It is also faster to toss a shovelful of concrete into a big interior space than to diligently stack small stones in layers of mortar to achieve the same end. Mortar is not as good when used in big chunks; it needs to be broken up with stones.

Mixing Concrete for Bedding Rubble. The easiest way I have found to mix a batch of weak concrete for bedding rubble is to mix the sand, cement, and water, without dry-mixing, in a big wheelbarrow. Slosh the mortar around until everything is mixed into a uniform, gray soup. Then slowly add the gravel until the mix is thick. It takes one experimental batch to determine the amount of water needed. I do this in a

big wheelbarrow with 4 shovels of portland cement and 12 to 16 shovels of sand. I can usually fit about 24 shovels of gravel.

The 1:3:6 or weaker concrete mix must be used pretty fast because it doesn't stay flexible for long. I usually get a course of faces all set, and then mix a batch of concrete to fill all the spaces quickly.

On a thin facing, where adhesion is important, the concrete bedding would not work. I have used it with success in thick facing jobs, as well as 2-foot-thick walls. Outside, in areas of extreme temperature swing, stones laid in concrete might not have the same ability to absorb shock without fracture as they would if laid in a limy mortar. Matters of this nature are devoid of statistics.

Outside Retaining Wall

FIGURE 8-6 shows a retaining wall of block being faced with stone. In this case the facing is 16 inches thick—more like a true stone wall, but a facing nevertheless because it is built against another wall. The block wall is 16 inches thick as well, and doubles as a set of stairs. The stone face is taller than the block, to give a curb to the retaining wall/steps. The total thickness of this wall is 32 inches, and it has proven worthy of the task of holding back 8 feet of dirt for five years.

The pointing mortar in this wall remains without cracks even though the wall was laid in concrete. The large size of the aggregate is not enough to prevent a tight fit between stones. Most of the tight-fitting look in a rubble wall, if it has it, is a phenomenon at the surface of the wall. The stones often get smaller as they go in to the wall, leaving larger vacancies in the wall than would seem likely looking at the face. You might find this technique faster and less messy than the conventional soft mortar. It is definitely cheaper and has the advantage of setting quicker so that you can build higher in a given session without crushing the mortar below.

Notice that the ends of the wall in FIG. 8-6 are built up higher than the middle. The stones are in crude courses. The long stone going up the ramp will be used to cross several smaller stones, whose tops form a flat surface to accept the flat bottom of the big stone. The white and gray splotches in the upper right are patches of leftover mortar from another job. Instead of throwing it out, I parged a little of the wall. The rough texture of the block is good for holding mortar. Also, the rough texture of the concrete between stones gives the pointing mortar something to grab.

At some point in thinness, a wall or facing could not be laid in a separate bedding mix. There wouldn't be enough space to keep the mud recessed for the pointing mortar, especially if you wanted the finish mortar surface somewhat recessed. Eight inches is something of a minimum thickness for bedding and pointing separately.

There are many approaches to laying and pointing stonework. Rules don't really exist, and when you

Fig. 8-6. A thick structural facing is laid against a block retaining wall. The stone here is bedded in concrete.

try to use them, they get broken. I knew nothing about it when I started. I have learned from doing, but am still experimenting. I don't think I've ever used the same mortar mix twice on different jobs. The type of job has so much to do with the approach.

I have read about how the stonemasonry was done at the Capitol Building in Washington, D.C. It is no problem for the masons to abide by the incredibly persnickety standards set by the government. I've rooted through old literature to gain glimpses of how the huge stone cathedrals and castles in Europe were built. As long as the king, the United States Government, or the Roman Catholic Church is picking up the tab, it is crazy for a builder to suggest anything but the finest way to do something. If it is your money and you have a lot of it, naturally as a builder, I will suggest the most conservative, heavy-duty way like it is the law. That way, I won't have to worry about my name getting trashed by an experimental footing that wasn't strong enough.

These situations are hardly normal in everyday construction, however. The first thing customers tell me is that they are broke. Always. Because I too am almost always broke, I have learned ways of making things with less money for myself. When clients sadly tell me they can't afford the wall they want me to build, I can't help but mention some of the money-saving options. On inside jobs, leaving a wall unpointed by design is an option that saves a lot of time and money, without sacrificing aesthetics or safety.

THE DRY-LAID LOOK
IN A MORTARED FACING

FIGURE 8-7 shows an 8-inch stone facing, built against a wooden wall, but independent of it. The stone wall has its own footing, and I could see no reason to attach it mechanically in any way to the plastered wall, which was already finished when I came on the job. There is no way for the 8-inch wall to fall over. It was built with integrity in courses with a good, adhesive mortar mix. Attaching it to the wooden wall with brick ties or anything else, in hope of gaining more stability, would mean the stone wall would have to suffer all the little changes the wooden wall goes through. Having masonry structurally dependent on wood is a poor approach. If you really need the ties, glues, and forms to make your stone wall, it is probably too thin and lacking in integrity.

Notice the fire thimble in FIG. 8-7. The purposes of this narrow wall are many. First and foremost, the owner wanted something artsy made of something old. It is hard to beat stones for old. Local, random stones also impart an organic feel to an otherwise overly sterile and uniform environment. Second, the owner wanted the look, or the suggestion, of a chimney. Third, he wanted greater fire protection from the wood stove that now sits in front of this interior chimney facade. Last, the two tons of stone manage to store some of the heat from the stove after the fire is out. The stone wasn't actually necessary to meet fire regulations, and both for this reason and for ease

Fig. 8-7. This facing gives the feel of a stone chimney for a wood stove.

of construction, it was built in contact with the plaster-board wall, instead of on standouts with an air space between walls.

To give you a feel for the possible procedure on a job like this, I kept a photo series of the whole four-day job. I gathered three, one-ton loads of stone in order to have extra to pick through. The actual volume of the wall (10 × 3 × 2/3 feet) is only 20 cubic feet. I used three bags of black masonry cement and less than a ton of sand. The thimble was in place before I took on the job.

Layout

The layout of a small interior job like this is easy and fast. The width of the structure was predetermined by the windows to either side. The fire thimble suggests centering in the wall. I struck chalk lines along plumb marks made on the wall itself, and then hung two plumb bobs in line with them, but out from the wall 8 inches. Those four lines described the solid object to be built. I checked sides by sighting across a plumb line to the chalk line on the wall. I checked faces by sighting across the twin plumb lines 8 inches from the white wall to see what stuck out and what stuck in. I set the corners with the points on the bobs.

FIGURE 8-8 shows the little wall after the first few courses are in. Notice the chalk line plumb bobs hanging near the corners. If you were to step around to the side of the work, you could sight along these two strings (when they weren't moving) to determine where the stones were to go. If the wall were plumb, you could just measure from the wall 8 inches. I placed the bottom left stone first, with its corner set to the plumb bob. I put in the lower right corner next, and then the face stones in between. Then I filled the rear gaps with mortar and junk stone, and used a cap of mortar to tie the course together, creating a new flat place to start again.

At no point in a wall this thin should you use out-sloping top surfaces. A stone can slope to the sides, unless it is too near the end of the wall, where it would tend to spill the stone above off the end. Also avoid in-sloping stones on a narrow wall like this, unless they are so shallow that their slope can be neutralized on the back side of the wall.

The black mortar is kept recessed in this facing. There is no jointing job. The stones are the dominant

visual element in this discount technique. See how the smaller stones are used to fill in the spaces that are left after the big stones are set. The long stone at the left end has a nice, flat top and bottom, and it crosses four joints below. This is why I had to overlook the fact that it has no real corner, but comes inward at less than 90 degrees. This type of concession is part of the deal in rubble work. You don't find ashlar stone lying in the field, free for the taking.

Notice all the joints are crossed. The stones are not smeared with mortar. Each new course begins with a stone that either overlaps the stone below or doesn't extend as far. It would be a mistake, visually and structurally, to begin a new course with a stone that was the same length as the one below. Reel in the plumb bobs slowly as the wall goes up. A tiny pair of vise grips clipped on a string makes a real handy plumb bob.

Black mortar works much better than lighter mud to achieve the dry-laid look. On casual observance, it is hard to tell the difference between a shadow and the black cement. The rougher the shape of the stone, the more obscure the courses of stonework become. There is just the right amount of wildness in some walls that keep them from being boring like block, or structurally inadequate like flagstone facing. I have noticed over the years that people like the wild, curving places in the stonework, and especially the way stones with odd angles fit together. This is never what I'm trying to do. I'm trying to build a boring block wall, but the crazy-shaped rubble stones won't let me.

FIGURE 8-8 represents a point in the progress where the mason says "Phew! I'm back on track again. That one almost got away from me." If you actually tried to build in the wild phenomenon, a wall could become subtly menacing, instead of soothing and protective. I add whimsy and lightheartedness to stone walls with nonstructural elements, such as the use of different colored stones, an occasional fossil stone, or an oddly shaped stone that is of special interest. I like to sneak little surprises in the spaces between poorly fitting faces. (See Chapter 13.) The old masons frown on this fooling around, but stonemasonry isn't serving the same purpose as it did way back. As I have claimed earlier, most of the stonemasonry going on these days is strictly for show. Let the show go on.

Fig. 8-8. The bottom couple courses are brought to a level surface.

Fig. 8-9. The ends are built up first with facings too.

FIGURE 8-9 shows the wall brought up several more courses. At the top, obviously the ends have been put on first. The middle is coming up to meet that height, in layers, with small stones actually laid flat, like bricks. There is a cavity behind the top middle stone, which doesn't quite reach the wall. I added some fill to that space. Then I brought the area level with a smearing of mortar. Black cement is expensive, about $11.00 per bag. It could be worthwhile on this type of job to have a batch of regular cheap mortar on hand to fill these cavities in the back of the wall, where it obviously won't be seen. I could have saved at least $10.00 by using cheap mortar on this job.

FIGURE 8-10 shows another jump in the wall progress. You can see where thin stones have been used to bring stones to a level course. It is much easier to do this than to try to find stones with crooked bottoms that happen to follow the contour of the inconsistency you've got going.

Near the top of this illustration is an example of breaking the rules. The large angular stone was laid before the smaller stack of stones at the corner. The big angled stone was too good not to use. I wanted to make sure it got in the wall. Because it doesn't have a square end on either side, it had to be used in the middle. The stone was positioned strategically to break the uncrossed joints to the right of center below it. The small stack of stones on the corner isn't really kosher, but the stones are flat and stable, and the goof is covered nicely on top with a long stone. In other words, I wiggled my way around that hassle and got back to a flat surface again, to begin another course. The area is stable and more interesting to the eye than a boxier area.

Use Big Stones Where Needed

If you go back to look at FIG. 8-7, you can see how the top of the wall as it was in FIG. 8-10 was dealt with. I managed to find a stone that was exactly 8 inches deep with a very flat top and bottom. I split it in two with hammer and chisel and got two great caps for this wall. I also split the stone so I could get it up on the wall. The slab was very useful to cover all the oddities below and give a fresh start to the wall where it is needed. The thimble usually can be expected to cause some trouble with the rubble. It is like starting a course in the middle with a round stone. As it turned out, I slipped around the chimney thimble without disrupting the normal masonry procedures. Above the thimble, I needed another long stone to bridge weight over the thimble itself, which is as delicate as a casserole dish. This was a great opportunity for me to make good use of the other half of the big stone I split. Because the top of this wall comes to a sloped ceiling, there was no point in me saving the big stone for a capstone. Above the thimble was probably my last chance to get the monster on the wall.

That sandstone was easy to split along a natural bedding plane. This type of stone cutting is instantly available to a novice. Just keep smacking a wedge-shaped chisel with a hammer along the line of a bedding plane in the stone.

Going Higher

This kind of job hardly warrants legitimate scaffolding, yet something must be done to stack stones above your reach. In the foreground of FIG. 8-7 is a makeshift scaffold with an oak board on cement blocks. Boards on blocks are the premier low scaffolding. Make sure the board is strong and the span isn't too much. The stone over the thimble weighs 150 pounds. A person lifting that on the board might be putting well over 300 pounds on the board. If it takes two people to lift a big stone, you can easily get 500 pounds on a board before you know what happened. In midlift of a heavy stone is an awful time for a board to break. Pay attention to what you're standing on. It is very easy to get scaffold and equipment overloaded in stonemasonry. It is nice to have a board of mortar, a bucket of shims and fill stone, and several face stones up on the board so you don't have to jump up and down constantly.

The top of FIG. 8-11 shows that the ceiling is angled This changes the strategy somewhat at the top of a wall. You need to look for stones that have the approximate angle of the ceiling on their tops and that are pretty large. About two courses short of the top, you need to check out the best stones you have left,

Fig. 8-10. The rules are broken to accommodate the big angled stone.

130

and measure to see what will fit. All too often, the top of a wall shows that the mason was running short of good stones and enthusiasm. That is unfortunate because the eye is drawn to the top of the wall. The approach of rubble stonemasonry, especially when gathering stone, is to mimic ashlar masonry as much as you can, and when you can't (which you can't), you fall back on a bunch of tricks, like avoiding out-sloping tops, shimming backs to make tops level, filling between in-sloping stones in layers, and so on. It is always a bit of a strain to make a rubble wall. There are many decisions to make with very vague procedures.

FACING FOR FIRE SAFETY

Many people want to back up their wood stoves with an inflammable wall of stone, to capture some of the feel of a stone fireplace. The previous project had a thimble going through the wall in a protected area, and needed no additional clearance from the wall. The wood stove will need to be far enough away from that narrow wall to maintain proper distance from the sides and the frame wall behind it.

FIGURE 8-12 shows the beginning of a very thick facing, which is actually a 2-foot-thick wall built against a sheet of plywood. The plywood is temporary, and spaced from the interior house walls. This wall needs the air space behind it because it is serving as a chimney. If it had been built in contact with the frame wall,

the frame could get dangerously heated during a hot fire. This wall only goes to the ceiling, and then turns into a block chimney to save money. The exterior part of the chimney at this residence is so hard to see that it wasn't worth the extra expense to take the stone chimney through the ceiling and out the roof, higher than the peak of the roof.

As seen in FIG. 8-12, the plywood backing is nailed onto horizontal studs that protrude beyond the plywood. The 2 × 4s are nailed into vertical wall studs through these extended parts. The temporary wall was easy to take apart when the chimney was done. I removed the nails from the 2 × 4s and slid the 2 × 4s out. This left a freestanding wall 2 inches from the plasterboard surface. It would be nearly impossible to form the back face of a wall that was within 2 inches of another wall without a form to build against. It couldn't be pointed. The nails holding the plywood to the standoff studs are pulled out as the wall is brought up. I used two-headed nails for this operation.

Near the top of an inside wall is a very awkward area to move big stones. The ceiling won't allow you

Fig. 8-12. This wall is built off the plywood backing. The wood will be removed, leaving a 2-inch space between the chimney and the plasterboard wall.

Fig. 8-11. Walls that don't end level need angled capstones with flat bottoms.

to get your body high enough to lift well. It is nice to have a helper in situations like this. In FIG. 8-13, we are test-fitting a peculiar rock with a curved bottom to see if it will fit over the hump on the stone below it. There is really no way to measure these irregular fits. You often need to lift the heavy beasts an extra time to see if they are going to work. In this case it did fit, so one person tipped the stone up while the other got a good glob of mortar on the new bed. The stone was then mushed into the mortar heap and set to the correct position. The sides and the back of the stone were packed tightly full of mortar. In this case, the stone wall was to have a conventional pointing job, so the bedding mortar was placed a few inches from the face plane.

FIGURE 8-14 shows the stone laying completed and the pointing job begun at the top. This wall used less than a bag of black cement for the pointing. Because the back half of the chimney does not show,

it was much faster to build than a two-sided wall. The effort to make stones look good next to each other represents much of the time-consuming part of stonemasonry. It is definitely a lot faster if you aren't concerned with the faces of stones.

This project is perhaps more practical than the previous one because it contains the chimney flue, whereas the previous project already had a block chimney ready to go without the stone facade. This block chimney would get a stone facing.

FACING A BLOCK CHIMNEY

Stone facings are commonly done on block chimneys. With the advent of blocks made to surround flue-liner sections, people gained the option of building a masonry chimney in one day. Many people build these quick chimneys with the notion of doing a stone facade later. It is not a bad idea, if you need to put a stove in quickly and you don't want to wait while a

Fig. 8-13. A curved stone is dry-tested for fit over a humped stone.

132

stone chimney is being built. The dilemma here is that you need to make a much fatter chimney when you put 8 inches of stone on the block than if you put the 8 inches of stone around the flue liner.

When you face block chimneys, in a sense you are making the blocks obsolete. When you finish the chimney facing and realize the blocks were not needed, you will want your money back.

The only advantage of doing a stone chimney in two steps—block chimney, then stone facing—is to gain some time to do an unhurried facing after the big construction rush is over. Another possible reason to do a stone on block chimney is if you are facing a block chimney that was never intended to be anything but a block chimney, such as when you want to snazz up your new house a little. One more possibility is that you want to give stonework a try, but you are nervous about actually building the chimney. In that case, you could hire a mason to build the basic block and flue-liner chimney. Then you would have a chimney you could use while you went about facing the chimney.

It is much less intimidating to merely measure off the block wall to set your stones than to set clay tile liners and build stonework around them. It is more like facing a block wall than building a chimney and is comfortably within range of the beginner.

FIGURE 8-15 shows a block chimney getting a stone facing. This chimney was a few inches off plumb, so instead of following its face with a measuring stick to set the stones, I made a template of the dimensions that I wanted the final chimney to be and set it on the top of the flue liner. It is held in place with the weight of a block on top. In each corner of the square piece of plywood, I cut a little notch to hold the string. The notches were all the same depth. The lines of the plumb bobs that hang from each corner are hooked on a nail in the top of the plywood. It is hard to see the lines in the photograph, but they represent the outer dimension of the facing, which will be plumb overall, even if the block chimney isn't.

If the bobs move around, you will need to make marks on cornerstones that you know are plumb with the top corner. Then you can hold the line against that spot and check your stones. You can check all sides of faces by sighting across the parallel lines.

The advantage of sighting over using an actual string to set stones to is that you can tolerate an oc-

casional protrusion out of the face plane. As long as you are aware that the stone has broken the line. When you use strings, however, no stone can come beyond the line or it will move the string and throw off all your references. When a stone penetrates the sight line, it doesn't render the method useless, unless you get so many rocks sticking out that you can't see past them to check other stones. The chimney shown in this series doesn't have a genuine 90-degree angle in it. For rubble corners, you must constantly split the difference between the direction of the blunder.

The chimney in FIG. 8-15 has metal connectors set in the mortar joints between blocks. This is a good idea because it will connect stone and block in many places and help them to act as one, more massive, unit. Connect the stone to the chimney, but do not connect the chimney to the house. Abut the side stones to the wall and caulk the joint between wood and stone when the job is done. It makes sense to wait for the mortar to cure, at which time it will naturally separate from the wood enough to make a caulk seal.

It is worth mentioning that a chimney needs a very good footing. There is probably nothing in a residence that puts more weight on the subsoil than a big, tall chimney. The one shown in FIG. 8-15 will weigh upwards of 10 tons when it is done. Those 20,000 pounds should have many square inches of something very solid to rest on. A chimney footing is no place to skimp. Plan it to extend at least 6 inches beyond the facing in all directions. If the block chimney you plan to face doesn't have that much footing, do some digging and add more concrete. The chimney shown here is roughly 3 feet × 3 feet. If the footing was 4 feet × 4 feet, you would have 16 square feet to spread the 10 tons over, or about 8 1/2 psi. You would also have a 6-inch extension of footing the whole way around the facing.

FIGURE 8-16 shows the chimney facing near the top. At the middle left, you can see a stone that slopes inward quite a lot. The depression in the wall will be leveled, ready for another stone the height of the brick shape cornerstone at the top left. I usually rotate which corner starts a course. You need to evenly distribute the good cornerstones so the thing doesn't look ashlar on one side and Neanderthal on the other. One of the most common blunders of the novice is to reach

Fig. 8-14. The pointing begins at the top of the finished chimney.

Fig. 8-15. Plumb lines can be hung from the top for a chimney facing.

Fig. 8-16. The 8-inch facing near the top of the chimney.

for the best stones right away and be left with stones that can hardly be stacked without those few good ones to cover the blunders.

FIGURE 8-17 shows the chimney near completion. This facing has a legitimate pointing job for maximum strength and weather protection. At the top of the chimney, the pointing mortar is still rough. It hasn't been *dressed*, or scraped back smooth and uniform. The plywood is still in place to protect the cap of strong mortar that connects the block and stone on top. The cap slopes away from the flue liner, which protrudes at least 2 inches beyond the block. The stone facing, then, must not go quite as high as the block because of the sloped out cap.

I pointed this facing as I built because I thought that there wouldn't be enough slack to leave the bedding mix recessed 2 inches. When you only have 8 inches of bearing surface, it gets trickier to leave the mortar tucked back. As it turned out, these stones were good enough that I could have done it in two steps and saved $30 and some mess. Black cement

Fig. 8-17. The mortar at the top of the completed chimney has yet to be dressed down.

is pretty messy to work with, and the less of it you need, the less messy.

Notice the scaffold in the foreground of FIG. 8-17. At this height, I was working at the top of the second unit. You should set scaffolding on blocks or boards that are set on a patch of firm, level dirt. The board or block is needed as a footing for the skinny scaffold legs. You don't want the pipe to sink in the dirt when you are on top. The bucks need to be set level. Any deviation will be exaggerated by subsequent units until you've got a leaning tower. It is worth the time spent to get your first unit set right. You can then put together the ones above without further adjustment. If you are only going to the height of one section, it isn't as important to have things level.

When you are doing stonework, it is not hard to overstock the scaffolding since you want to have enough stuff up there so that you can keep working for awhile without climbing up and down. Before you know it, you've got a ton of stuff up there. This can be a problem, in chimney work especially, because a chimney is so narrow it only takes one unit of scaffolding to cover its width. This means you have a tower of scaffolding, which is less stable than a wall of scaffolding and more prone to tipping over from being top heavy. To give the tower stability, put some heavy planks across the lowest bars at the bottom, and load the boards with stone, blocks, barrels of water, or anything that is stable and heavy to give the tower a lower center of gravity.

Another potential danger is to work with materials that are too heavy for the boards. It is not at all uncommon to catch yourself standing in the center of a 2-inch-thick board with another person, trying to lift a stone that weighs 200 pounds. If you are big people, there is already close to 600 pounds on there, and it wouldn't be unusual to have a mortar board with 80 pounds of mortar on there, too. It is funny that people will do this 20 feet up in the air, with big stones strewn about on the ground below. I've had some unpleasant rides to the ground on overloaded boards. Where do you think all the rocks and cement that were on the broken plank are headed?

Wide metal boards (FIG. 8-17) are better for putting a good load on. Most rental places carry scaffolding and the metal boards. They also have booms and pulley wheels for getting supplies up with a rope and a helper. You might be able to rent scaffolding from

135

a mason. On this chimney-facing job, it took one man 98 hours and used 10 tons of materials, the scaffolding only cost $26.00. Don't try to do a job of this nature on ladders or makeshift scaffolding.

The top 3 feet of this chimney overhangs the roof on one side by 8 inches. It is debatable if this much stone needs support other than the roof rafter and decking at the edge of the roof. It has an interior stone facing directly below to rest on if you don't care about the missing ceiling insulation. Without adequate support from below, the suspended stonework would need to be bridged to the two sides with a lintel stone or some hidden angle iron.

Chimney jobs are definitely less hassle with two people. It is hard to get stones to the top of a scaffold without a helper. On this job, I pulled all the materials up in a 5-gallon bucket with a rope on the handle. Of the 98 hours spent on this job, I must have spent 2 just climbing up and down the scaffolding for things I'd forgotten or dropped.

The courses in this facing were done just like all the previous projects, except there are corners instead of ends of walls. Having four short faces tied together involves a constant vigil to see which side needs the most help structurally and visually. On each course, the favored side should change, to keep things well integrated. These mountain sandstone were chosen for their bright and varied colors, and for the abundance of naturally occurring right angles. A chimney like this is nearly all cornerstones. The gathering part of the job is quite fussy compared to other stonework. No stone can be more than 8 inches deep, and the stones need to be shaped well to compensate for the skinny walls.

I use a slightly richer mortar mix for a stone facing without full wall thickness. On this chimney, I used one part type N black masonry cement to 2 1/2 parts sand, and mixed it to a creamy texture. When laying stone against porous block or concrete, it is important to keep the wall moist, or it will steal the moisture from your mortar before it has had a chance to set. The strength and adhesiveness of mortar in a tall chimney is much more significant than it is in the thick walls of a building. Don't let leaves and junk get in the sand, since they will cause breaks in the bond of the mortar.

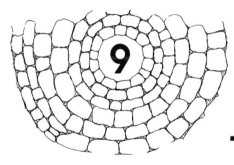

9

CHIMNEYS AND FIREPLACES

MASONRY structures that contain combustion must be built to a completely different set of criteria than other masonry. I caution you, the homeowner, from having a casual attitude about designing your own system for burning wood. Nearly anyone can build a successful low retaining wall; but building a fireplace and chimney, or even installing a prefabricated fireplace and chimney, is something to do by the book. Fire codes are designed to keep you from burning your house down. Many of the codes are getting stricter with regards to burning wood. This is good, in view of the statistics on accidental fires caused by wood-burning devices (TABLE 9-1).

There are a lot of places where you can go wrong when building a chimney for a fireplace or woodstove. Frankly, I don't recommend that the casual builder take on such a project. A more serious builder, however, will find enough information in this chapter to build a fireplace and chimney.

Chimneys are certainly easier to build with bricks than stone, and can be done smaller. Rubble stone needs a thicker wall, which scales up the outer dimensions of a chimney. Chimneys of rubble stone should have 12-inch-thick walls, whereas brick chimneys can be as narrow as 4 inches for one story or 8 inches for two stories.

WOOD BURNING

Wood-burning technology is still in its infancy. The large majority of appliances (stoves and fireplaces) now in use will fail pollution standards soon to be set by the Environmental Protection Agency. The state of Oregon has already set tough new pollution standards for wood stoves, which make obsolete nearly all of the models available ten years ago. The rest of the country will likely follow suit. Now that there are good alternatives, there is no reason for us as a society to put up with pollution-belching wood burners. New fire regulations being proposed will squeeze the old technology from another side. Pollution regulations and fire codes are rightly chipping away at an obsolete and dangerous technology. In the distant future, even the state-of-the-art woodstoves of today will seem shamefully wasteful of something as precious as a tree. Tighter houses with better insulation and solar help won't need major heating devices.

As of late 1986, the very best stove on the market, according to tests carried out by the Oregon Department of Environmental Quality, has an overall efficiency rating of 81.6 percent, meaning it is losing only 18.4 percent of the available heat from the wood. The losses are a combination of incomplete combustion and heat loss up the chimney. None of these tests actually accounts for the strain that a woodstove puts on the home's heat loss. For instance, a chimney will lose heat even when there is no fire.

Most chimneys are metal or masonry fingers, poking through the insulative barriers of a house. When there is a fire, the fire needs air—your air. You will have to let more cold air into the house to feed the fire the air it needs. Burning wood, or anything, inside a tight house is a health hazard, unless adequate ventilation is provided.

For residential wood heat, nearly all building codes now call for either an Underwriter's Laboratories listed prefabricated metal chimney or a masonry chimney with a fireclay flue liner. The brick, block, or stone of a masonry chimney is mostly a facing. The liners alone would act as a chimney, but they need more stability and protection. Rubble stone is not fit for actually creating a firebox or chimney opening.

Table 9-1. Injuries Related
to Wood-Burning Appliances.

YEAR	FIRES	PERCENT CHANGE	DEATHS	DOLLAR LOSS
1978	66,800		250	$134 million
1979	70,700	+ 14	210	$175 million
1980	112,000	+ 58	350	—
1981	130,100	+ 16	290	$265 million
1982	139,800	+ 7	250	$257 million
1983	140,600	+ 0.6	280	$296 million

Source: U.S. Consumer Product Safety Commission

Stone is not able to retain its integrity through the many expansion/contraction cycles endured by a fireplace. Fireclay brick is designed for that purpose. In a chimney, stone creates a rough interior surface, which encourages accumulation of soot and creosote. When the stuff finally catches fire, it is too hot for the stones and mortar of the chimney. A bad creosote fire, with temperatures sometimes reaching 2000°F, is too hot for almost any common container.

What is a Stone Fireplace?

A stone fireplace is an antique. I have seen several of them, built entirely of stone, and very large. These monsters were built where firebrick was unavailable, and you wouldn't want one as a heater any more than you'd want a horse and buggy to commute to work. What passes as a stone fireplace these days is actually a firebrick fireplace with a stone facing. What we call a stone chimney now is a clay tile chimney with a wall of stone around it. The firebrick and flue liners would function without the stone or brick facing if you could protect the delicate arrangement. In other words, there is no such thing as a stone fireplace or chimney anymore, and it's a good thing because stone is not good for that use, except as an outer layer, or facing.

The Paradox of Burning Wood

Before discussing chimney and fireplace design criteria, let's explore the problems of burning wood. For heating, cooking, and lighting, wood burning began in open pits. In early houses, a hole was left in the roof to let out smoke. This arrangement caused some very smoky houses, while capturing a very small percentage of the wood's available Btus. The next big

jump came with a hood over the pit that directed the smoke up a chimney and out of the dwelling. This helped the smokiness a lot, but not the heating efficiency. Many methods were developed to contain the heat longer on its way up the chimney. The experiment continues today, with perhaps greater fervor, yielding some interesting results.

When wood is burned at typical stove or fireplace temperatures, gasses are driven off without burning. These gasses contain as much heat energy as the wood that burns, but they are lost to the atmosphere, except for the part that condenses on the inside of the chimney on the way up. Typically, the gasses are toxic and degrading to the environment, and the part that remains to coat the chimney, called *creosote*, constitutes most of the danger of wood burning. When creosote builds up in the flue, it can be ignited by an exceptionally hot fire. When the soot starts to burn, it can cause a frightening, roaring blaze, accelerated by the increased draft it causes. The temperature of the new fire in the flue can reach 2000°F, which is enough to destroy a chimney and cause a house to catch fire.

It is because of the likelihood of a creosote fire that the codes are so seemingly strict. If there were no danger of this secondary fire happening, you could use a chimney of mud and sticks. The paradox is that the codes are too lax to protect from creosote fires. The fires keep happening, destroying lives and property. Government tests on different approved chimneys, where a creosote fire was intentionally brought on, showed outside surface temperatures much too high for the 2-inch clearance called for in the installation. The best of the prefabricated chimneys tested got 278°F on its outside surface during a creosote fire. Some reached over 700°F on the outside. Would you want a pipe that hot to be within 2 inches of dry pine framing lumber?

Wood can be burned at higher temperatures, and the gases are burned as well. You can get more heat from the same amount of wood, and decrease the emissions drastically. Decreased emissions from the fire also mean decreased buildup of creosote and decreased danger of burning down the house.

The problem with having a very hot fire is that it is more difficult to contain and to recover the great heat that escapes up the chimney. A variety of methods attempt, with varying degrees of success, to capture that heat before it escapes. Stoves and fireplaces come with baffles to reroute hot gases through radiators. Water jackets in the fire try to put some of the escaping heat into a tank of hot water.

BREAKTHROUGHS

The problem with most approaches to getting extra heat from the flue gases is that they slow down the escaping exhaust and cool it off. This makes the fire burn less efficiently, produce more creosote, and leave more of it deposited on the flue liner. The stoves and fireplace inserts I have seen with water jackets or cool-air grates directly over the flames seem to be the worst offenders for creosote accumulation. The coil of cold water over the fire invites the smoke to condense. The central paradox, again, is that if you have a slow, controlled burn, you won't lose much heat up the chimney, but you will have an inefficient burn that pollutes the air and water, and creates a considerable fire hazard. If you burn a very hot fire, the combustion is much more complete, but the extra heat is lost up the flue. Attempts at recovering the heat inevitably mean cooling the smoke and impeding the flow of the escaping gases. There have been a handful of notable breakthroughs, however.

Russian Fireplace

A *Russian fireplace* is a massive masonry structure with a firebrick combustion chamber designed for high-temperature fires. The flue is designed to carry the hot, already burned, exhaust through a series of brick baffles, where some of the heat is given up to the mass and stored. This kind of heater reportedly works very well, but is tricky to build because of the baffled flue. I wouldn't attempt one with rubble stone. Another drawback of the heater is that you need to light a fire every time you want to use it. Still, a Russian fireplace and other designs using the same principle are a big jump in safety, pollution control, and heating efficiency, compared to a more conventional wood stove or fireplace.

Catalytic Combustors

Recently, perhaps the biggest breakthrough in wood-burning technology has been the discovery of the catalytic combustor. It was found that wood gases of the type that normally escape combustion would burn at a lower temperature than usual if the gas was passed through a ceramic honeycomb that has a platinum coating. The hows and the whys of this magic alludes me, but I know that the platinum acts as a catalyst; that is, it causes a reaction without actually taking part in it. Stoves that are painstakingly designed for the optimum use of the catalytic combustor can achieve very low emissions readings, and overall efficiencies of 80 percent . . . according to the extensive testing of the Oregon Department of Environmental Quality. The low emissions means more heat from the same wood, and much less of the dangerous creosote buildup in the flue. This new breed of stove can be 100 times cleaner than an old cast-iron air-tight stove on slow burn.

Catalytic combustors, when added to existing stoves, evidently don't work very well. The stove must be designed to use the device in order for the device to get much out of it. Two of the drawbacks of the catalytic stoves are high initial cost (the little platinum-coated combustors will add $200.00 to the cost of a stove) and some trickiness in getting the fire hot enough for the combustor to do its thing. The combustor burns at 500°F the gases that normally would need 1000°F to burn. The trouble is, the stove needs to get up to 500° before the gizmo does anything. Another problem is that you need a fan to move air past the combustor fast enough to extract much heat. These stoves might be likened to a ten-speed bike: when you first get one, it is awkward, but when you learn how it works, it is obviously a big improvement over a one-speed.

Noncatalytic High-Tech Stoves

There are many stoves and fireplaces on the market which achieve a high efficiency without catalytic combustors. Most of them use the principle of a hot, fast, clean burn, which they achieved through placement of extra air intakes at crucial places in the stove and through use of an insulated firebrick firebox, which maintains the 1000°F and higher temperatures required for a clean burn of wood. The challenge of this approach is to get the heat from the burned gases before the exhaust goes out the chimney. This is often solved through the use of a baffled exit for the gases where they give up heat to the metal. Additional heat is extracted with forced-air circulation through interior chambers of the stove. According to test results, these stoves, though a vast improvement over conventional ones, are not as clean or efficient as a well-tuned catalytic stove.

Pellet-Burning Stoves

Another recent breakthrough in the science of burning wood is the pellet-burning stove. This type of stove incorporates a specialized fuel—pellets of wood-industry by-products that are twice as dry as seasoned firewood—which is fed into the stove by a slowly turning auger. There are several big advantages in this approach. One, you let industry mess with getting the firewood—no stacking, splitting, sweeping up bark, and no fussing with the fire. The auger slowly turns and drops the dry pellets of wood scrap onto a very hot fire. Two, the fire can remain very small and yet be very hot. The pellets burn fast and complete because of their high surface area and extra dryness. Three, the fire never gets bogged down or smothered; there is no smoldering. The high heat is drawn off with air circulation and heat exchangers.

The advantage and disadvantage of this kind of stove is the same: the specialized fuel that is needed. Possibly a similar design will be able to handle sawdust, which is so much more universal than the pellets.

Using Your Dinosaur Better

You can make a regular old wood stove perform a lot better by adjusting your behavior somewhat. The type of wood burned will affect the combustion efficiency a lot. The worst possible polluting stove would be one that burns a big hunk of green, wet pine in an airtight stove with lots of bends and elbows on a pipe that isn't quite high enough. Seasoned hardwoods contain less moisture and less pitch and will burn much cleaner than green pine. Using a straight-shot pipe with ample altitude will keep the exhaust moving fast. Having insulated pipe where it enters the outside will help to keep the flue hot, which will help the draft and lessen the amount of creosote deposited. If vents and dampers are left open, the combustion will be hot enough to remain fairly clean, especially if your dry hardwood is in small pieces. True, a good portion of your heat will be lost up the

chimney, but the increased efficiency of combustion should make up for this loss, especially if you can add mass to your stove.

When we had a baby crawling around, I wanted to make sure she wouldn't be burned by our cast-iron wood stove. I stacked bricks around the stove, without mortar, until the entire surface was covered except the door. When the stove was cranking, those bricks never got too hot to touch, even though they were in contact with the cast iron. The bricks added 500 pounds to the stove, so I made sure they were underneath as well, to brace the legs. The bricks kept the area near the stove from getting uncomfortably hot, and it carried the warmth long after the fire was out.

These kinds of adjustments won't make your stove as clean as the previously mentioned ones, but they will be a big help.

Wood Stove with Gas Heater

I think one of the main reasons people like to damp down a fire is to have it hold through the night. If the mass would hold the heat for the night and you could start another fire easily in the morning, I think there would be a lot less pollution from wood stoves. For this reason, I recommend a new type of stove that is coupled with a gas burner. The heater functions as a gas heater if you are out of firewood, or it functions as a wood stove.

The best part is that it will run the gas part long enough to catch the wood on fire, and then shut off the gas. This means that, for a few dollars a year worth of propane, you can stop worrying about fires going out at night or any other time. You can forget about having a box of dry kindling; you won't need to have a box of newspaper on hand; and if your logs are wet, you can run the gas jets a bit longer to get the wood burning. This type of hybrid stove is made in the catalytic mode. Some also use coal, which adds considerable versatility. The stoves can be opened for an old-fashioned fireplace effect.

MASONRY FIREPLACES

Masonry fireplaces have a lot of strikes against them, which I feel compelled to reveal before divulging more information about their construction. Strike one is burning wood in general. Even when used at its highest efficiency, the wood habit implies buying the wood, or owning your own and cutting it. Cutting it implies using a chainsaw and a truck. In either case, you'll need a place to stack it, near the house, but out of the rain. You'll need to stack it and restack it, carrying each piece many times before you finally haul it out of the stove as ashes, and scrape it out of the chimney as creosote. It is a lot of hassle for the heat, and it is dangerous heat that can burn down your house.

Another drawback of burning wood is the way it wrecks the piece of a forest. City dwellers are often surprised to find the quiet countryside roaring with chainsaws. The chainsaws are dangerous themselves, and create more dangers, especially if you are felling large trees.

As you can tell, I'm not a big advocate of woodburning. Passive solar heat makes a state-of-the-art wood burner look archaic. Superinsulated houses are heated by the appliances and the residents. Still, there is a place for burning wood, and the special atmosphere it creates. Even in special circumstances, however, it is hard to advocate a masonry fireplace with all the beautiful, efficient stoves available that allow you to view the fire. The stoves are faster, cheaper, safer, and cleaner, and you will get much more heat from our precious wood.

If you are still compelled to have that special appeal of a masonry fireplace, you should know this: thousands of people have decided to improve the performance of their masonry fireplace by adding an insert or by using a wood stove that goes in the old fireplace opening. FIGURE 9-1 shows a side view of a typical fireplace insert. FIGURE 9-2 shows a side view of a wood stove adapted to an existing fireplace.

There are dozens of fireplace conversions on the market, all designed to improve the performance of a masonry fireplace. The tragedy or irony is that the fireplace inserts render obsolete the difficult and exacting masonry required for a masonry fireplace.

Prefabricated metal fireplaces, on the other hand, already have the correct proportions and are designed to be covered with a masonry facing, which is a task much more likely to be mastered by the handyman. If you use one of these fireplaces, you can have the same effect as a masonry fireplace, often with better results, and you don't need to know how to build a fireplace.

As you will see, there are a lot of places you could mess up with a masonry fireplace. Using rubble stone makes it more difficult. The sizes and proportions of the various surfaces in a masonry fireplace are crucial to its proper functioning. If you do it wrong, it will be difficult to correct.

It is a waste of effort to build a masonry fireplace if you think that someday you'll put a wood stove in front of it, or put an insert in, like so many do. Before you begin to build a masonry fireplace, consider these options:

☐ Change to solar heat or superinsulation, or move to a warm place.
☐ Build an outside insulated wood furnace that brings the heat indoors through forced air or water, thereby eliminating the indoor mess, the oxygen depletion, the draftiness, and the fire hazard.
☐ Buy one of the new catalytic or high-tech stoves and a prefabricated stainless steel chimney.
☐ Buy a prefabricated, efficient steel fireplace, and build your masonry around it, according to manufacturer's instructions.
☐ Build a Russian fireplace with brick. (as explained in *Fine Homebuilding Construction Techniques,* The Taunton Press, Inc., Newtown, Ct.).

A STONEMASONRY CHIMNEY

The purpose of a chimney is to safely carry combustion by-products out of the house. A chimney must do this fast enough to keep the fire going. As I've said, rubble stone is not a fit material to create a safe chimney, however, it is fine, though bulky, for building the protective walls around the fireclay flue liner, which is the actual chimney. Stones without the liner, although they have certainly been used, make too rough a surface and cause a drag on the escaping gasses. The rough texture of the rocks also gives creosote nice places to condense and form the dreaded deposits. Stone also tends to self-destruct after repeated heating/cooling cycles.

The vitrified clay of a liner is designed for extreme temperature swings, although the liners do crack and bust in a bad chimney fire. The recommended,

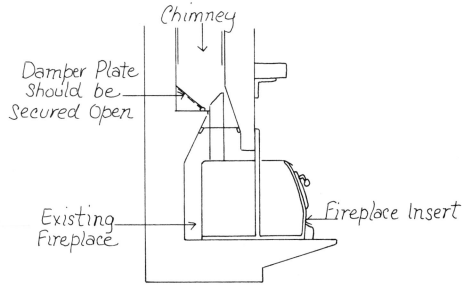

Fig. 9-1. A typical fireplace insert.

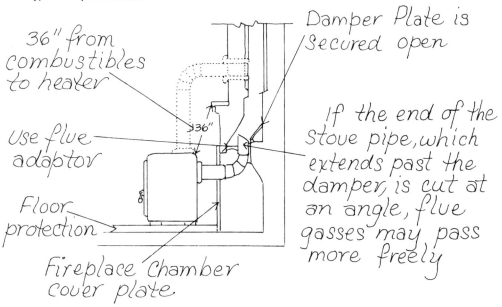

Fig. 9-2. A fireplace converted for wood stove use.

prefabricated metal chimneys also are destroyed in a real creosote fire. The big difference is the likelihood of damage to the house. In similar tests with creosote fires, insulated metal chimneys got hot enough on the outside of the pipe to catch wood on fire. Masonry chimneys reached an outside temper-

ature of 114°F during the soot fire. The great mass of the masonry keeps the thing from getting very hot during the combustion of a relatively small amount of creosote.

All the chimneys suffered damage in the test fire. The masonry chimney had broken liners, but didn't generate enough heat on its surface to burn down a house. The metal pipes tested all got hot enough to boil water on the outside surface of their insulated covering during a creosote fire. My only conclusion from all this is that creosote fires are unacceptable, and we must learn how to prevent them from happening if we are to continue burning wood. Would you drive a car if you felt that there was a fair chance of the gas tank blowing up? We have learned how to use gasoline in relative safety. We can learn how to burn wood, too.

Building Codes

TABLE 9-2 is a listing of various building and fire codes prepared by the Tennessee Valley Authority (TVA). Sources and reasons are given for each design criterion. Local codes might overrule these codes, and, as indicated, some of the codes might become stricter soon. Failure to comply with these guidelines could make it tough to keep fire insurance, and for good reason. Oddly, future pollution control, which we need pretty badly unless we want to throw out the world, might make a masonry fireplace, even as done by the book, obsolete and unlawful to use.

Again, because the accumulation of creosote (and the possibility of it catching fire and burning in the chimney at temperatures too hot for the materials and the surroundings) constitutes the greatest danger in wood burning, the emphasis should be in removing and avoiding the accumulation of creosote. Burning very dry wood at high temperature helps a lot. Having a straight chimney with a slick, round inside surface that extends high enough into the open air will help reduce the accumulation. Regular inspection and chimney sweeping as needed are essential when burning wood, except possibly with some of the most advanced stoves. The most sensible approach to creosote that I've heard is to burn it off before it can form and get some heat out of it. This means a fire of 1000°F or hotter, except with a catalytic combustor, which will burn the stuff at 500°F.

Stone Chimney

FIGURE 9-3 shows a cross section of a stone chimney. The same type of chimney, done with bricks, could be considerably smaller. The footing should be of concrete 1 foot thick, beginning below the frost

depth on solid ground. It should extend beyond the outer chimney walls by at least 6 inches.

The flue liners must be laid in refractory cement mortar (ASTM C-105) in a very thin joint. A thin joint, if destroyed, will cause less damage. The flue liners must have a space around them so they can expand without breaking. You can use mineral wool to maintain the space for the slight expansion. With brick or block, that space is easier to maintain than it is with rubble.

At the top of the chimney, the gap must be closed and the walls of the chimney tied together with a cap. Don't confuse a chimney cap with a rain cap. The chimney cap is like a wall cap, whereas the rain cap keeps water and downdrafts out of the flue. The chimney cap should slope outward to shed water, overhang the chimney walls, and preferably have some sort of drip edge on the underside of the overhang (FIG. 9-3). The drip edge will keep water off the surface of the chimney. The flue liner should extend a minimum of 2 inches above the chimney walls.

The diameter or width of the flue is an important factor in determining how much smoke will be able to rise out. The altitude of the chimney, the surrounding objects, and the wind will have a lot of effect on the way a chimney draws. The resistance of the flue is also a factor. If the flue has bends, elbows, and a rough interior surface, the flow of air will be stodgy and soot will build up. Common sizes for clay flue liners are 8 × 8, 8 × 12, 12 × 12, and 12 × 16 inches. Round liners are better. They are stronger, with less resistance and less surface area for the volume of air they contain. They are harder to find, though, and come in the following diameters: 6, 8, 10, 12, and 15 inches.

Foundations for Chimneys

A chimney needs a sound footing. FIGURE 9-4 shows the suggested dimensions. It is important that a chimney footing not be built on fill dirt that will settle, even if it is below the frost line. A concrete footing for a residential chimney must set at least 24 hours before supporting more weight.

Walls and Flue Liners

Build up the outer walls and flue liners simultaneously. The flue liners must be laid plumb in a very thin

143

Table 9-2. Building Codes for Chimneys.

	CODES	REASONS	SOURCES
CONSTRUCTION TECHNIQUES	Masonry chimneys shall not be corbelled from a chimney wall that is less than 12″ in thickness.	A	all
	Chimney shall have near vertical alignment.	A,E,F	all
	Masonry shall be laid with full push-filled head-and-bed mortar joints.	A,I	all
	All masonry chimneys shall have separate clean-out openings equipped with tight-sealing iron doors for each flue.	C	all
FOUNDATION	The foundation must rest on solid (not filled) ground.	A	all
	The foundation can be poured especially for the chimney or can be part of a solid masonry wall.	A	all
	The foundation for the chimney must be at least 12″ thick.	A	1 & 2
	The foundation must extend a minimum of 6″ beyond the face of the masonry resting upon it.	A	1 & 2
	The bottom of the foundation must extend below the frostline.	A	1 & 2
CLEARANCES & FIRESTOP	An exterior chimney may have zero clearance at the sheathing and must have a 1/2″ minimum clearance to other combustible materials. (Note: Future codes may require greater clearance to exterior chimneys.)	H	NFPA
	An interior chimney should have a minimum 2″ clearance to combustibles.	H	TVA, NFPA
	All spaces between the chimney and the floors and ceilings through which it passes shall be firestopped with noncombustible materials.	H	all
	The firestopping between the chimney and wood joists, beams, or headers shall be galvanized steel not less than 26 gauge OR shall be noncombustible to a depth of not less than 1″.	H	all
	A change in the size or the shape of a chimney flue where the chimney passes through the roof shall not be made within a distance of 6″ above or below the roof joists or rafters.	A,B,H	all
	The minimum chimney height for "fire safety" is the greater of 3 feet above the highest point where the chimney penetrates the roof line OR 2 feet higher than any portion of the structure (or other structures) within 10 feet of the chimney.	D	all

KEY TO SOURCES

BIA	=	Brick Institute of America	TVA	=	a recommendation made by TVA in excess of the codes
BOCA	=	The Building Official Code Administration			
ICBO	=	International Conference of Building Officials	UBC	=	Uniform Building Code Standard
NFPA	=	National Fire Protection Association	all	=	all of the above
1 & 2	=	One and Two Family Dwelling Code			
SBCCI	=	Southern Building Code Congress International	NOTE:		Remember that local codes may supersede those shown above.

	CODES	REASONS	SOURCES
MATERIALS	Chimney shall have proper support consisting of masonry and reinforced concrete.	A	all
	Masonry chimney walls shall be constructed of:	A,H,I	all
	1) solid masonry units not less than 4″ (nominal) in thickness, OR		
	2) reinforced portland cement concrete or refractory cement concrete not less than 4″ (actual) in thickness, OR		
	3) rubble stone masonry (mountain stone) not less than 12″ (actual) in thickness		
	Modular concrete flue blocks shall be sealed against moisture with an appropriate material.	B	TVA,BIA
	Liner joints should be set in high temperature refractory cement. (TVA recommends a finished bed joint thickness of 1/16″ to 1/8″.)	B,I	all
	The thimble shall be of 5/8″ fireclay or metal, at least 3/16″ steel, OR 26 gauge stainless steel or the equivalent.	A	all
	Flashing shall be non-corrosive metal or the equivalent used to cover and protect against leakage where the roof decking approaches the chimney walls.	B,I	all
CHIMNEY FLUE	A masonry chimney shall be lined with either 5/8″ fireclay or the equivalent. (TVA considers 26 gauge stainless steel or a safety-listed factory-built chimney as equivalent.)	B,F,H,I	all
	Multiple connections to a single flue are not allowed.	D,E,G	all
	Where more than two flues are located in the same chimney, there shall be no more than two flues in any group of adjoining flues without wythe separation.	A	all
	The wythe shall be separated from the flue liners by an airspace.	B,H,I	all
	The wythe shall be 4″ (nominal) in thickness.	H	all

KEY TO REASONS

A. **Observing Basic Construction Rules.** Some chimney codes and guidelines are concerned with the basic construction of the chimney. These codes deal with the type of materials used and the techniques used in building the chimney.

B. **Preventing Deterioration and Weathering.** Other building and materials codes are designed to help prevent the deterioration of the chimney. Weather, high temperatures (in the fire and smoke), acids from incomplete combustion, and creosote deposits are the main causes of chimney deterioration.

C. **Ensuring Easy Chimney Maintenance.** Proper placement of a clean-out door makes chimney maintenance easy and increases the likelihood that the owner will take care of the chimney. A properly maintained chimney is a safer chimney.

D. **Reducing Downdraft.** Several codes seek to reduce the chances of "downdraft." Downdraft occurs when wind blows down the chimney or when the chimney is not constructed properly and smoke is forced back into the house.

E. **Reducing Creosote Accumulation.** In addition to fueling a hazardous chimney fire, creosote can prevent the chimney from functioning properly by slowing the smoke as it rises from the fire. Creosote also has chemical properties that contribute to chimney deterioration.

F. **Reducing Resistance in the Chimney.** One of the main purposes of a chimney is to conduct smoke away from the fire and out of the house. a number of construction Practices Like corbelling (bends) or a rough brick wall can create "resistance" in the chimney and prevent it from carrying out this important function.

G. **Controlling a Fire.** There are two ways of controlling a fire for proper burning: control the fuel supply and control the air supply. The owner can control the fuel supply by the number of logs fed to the wood heater. Controlling the supply of air that reaches the flames is done by air inlets on a wood heater. Controlled burning is important for safety as well as for comfort.

H. **Preventing Heat Transfer Through Solids.** One of the ways house fires start is from heat passing through a solid (such as a masonry chimney wall) and igniting a combustible material (such as a wood stud). Codes seek to eliminate this type of heat transfer by requiring clearance to combustibles.

I. **Preventing Mortar/Chimney Wall Cracks.** Another way that fires start is from hot smoke and air passing through cracks in deteriorating mortar or chimney walls and igniting nearby combustible materials. Codes also seek to prevent this from happening.

Courtesy Tennessee Valley Authority

145

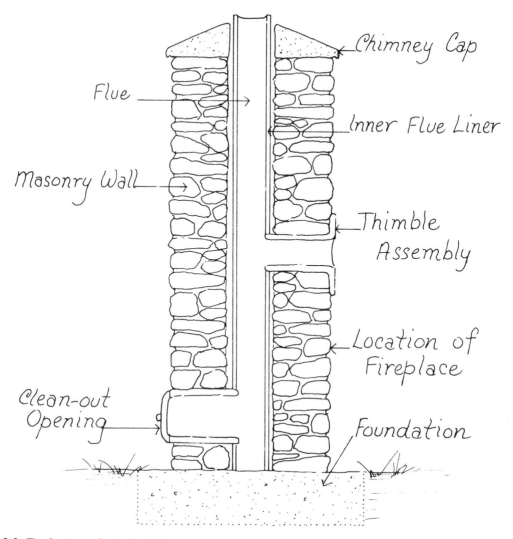

Fig. 9-3. The features of a typical chimney.

bed of refractory cement or fireclay mortar. After you lay the tile, reach inside and smooth the joint so you do not leave burrs of mortar to obstruct the flow. Smear the outside of the joint smooth and fill it completely. FIGURE 9-5 shows a typical liner and stone chimney getting started.

You do not need to lay the stone in fireclay mortar; just use this mortar for the liner, the inner hearth, and the combustion chamber if you build a complete fireplace. Type N mortar (One part masonry cement, three parts graded sand) should be fine for the chim-

ney walls. If the walls are narrow, a mortar with greater adhesion, such as type S, might be in order.

How High To Build?

Before getting to the various openings in a chimney, you should know how high to build your chimney. FIGURE 9-6 shows the standard code for the height of the chimney in relation to nearby objects, like the roof. A taller chimney is in order if your house is situated in a naturally stagnant area, or one with downdrafts.

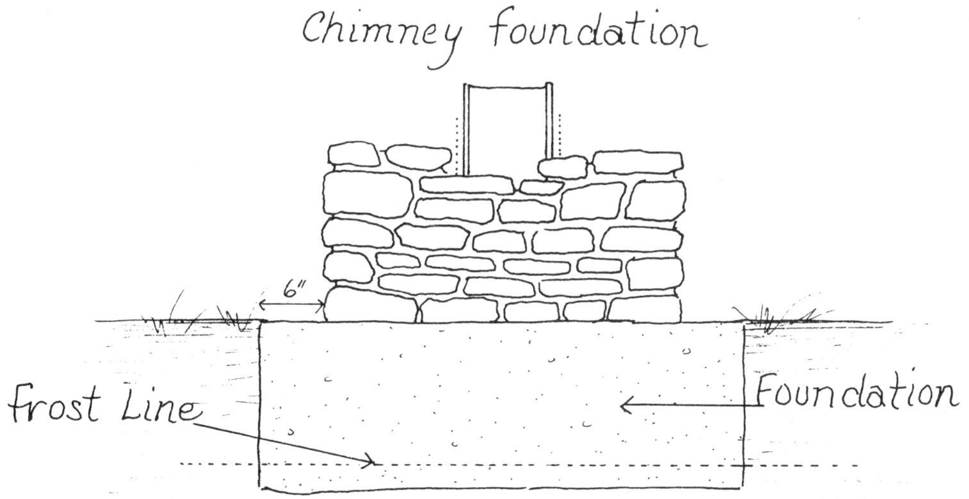

Chimney foundation

6"

Frost Line

Foundation

Fig. 9-4. *A chimney footing must be extra sound.*

Chimney Wall and Liner Construction

Fig. 9-5. *Stones are laid around the liner and spaces filled with small stones.*

Fireclay flue liners are not exact in dimension. It is hard to take an accurate course with one by trying to plumb the sides. I like to set them to a plumb bob when possible. If you see that the opening is centered with the string, you will be able to place the liner correctly and achieve an overall plumb flue.

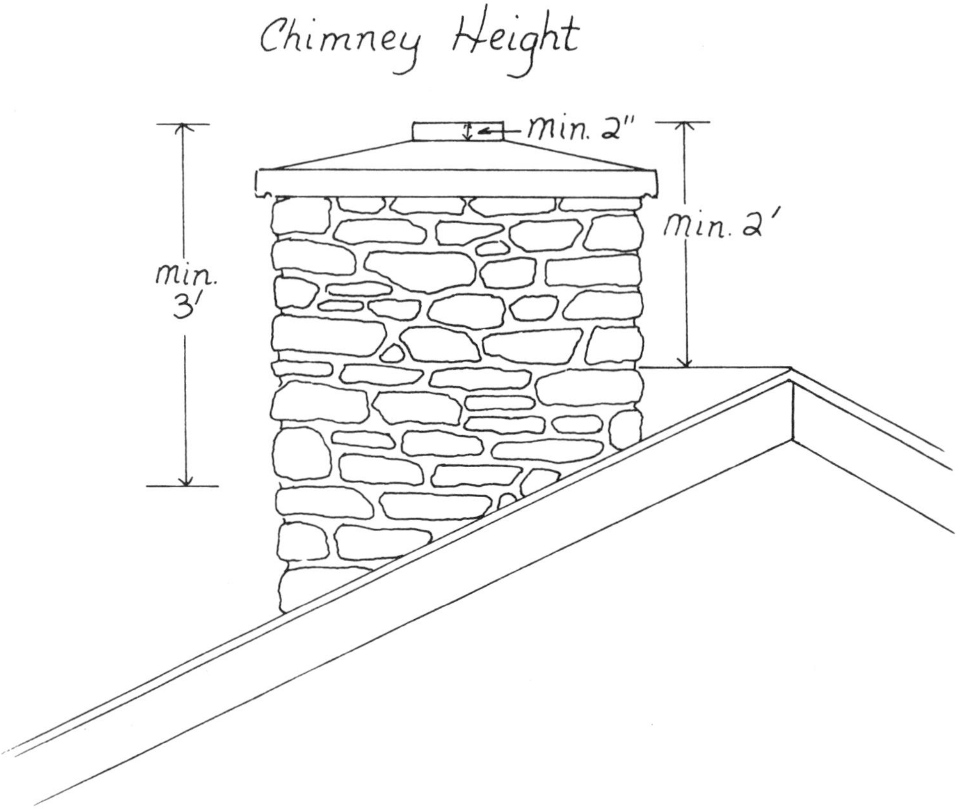

Chimney Height

min. 2"

min. 2'

min. 3'

Fig. 9-6. A chimney must be 2 feet higher than anything within 10 feet of it.

FIGURE 9-7 shows a chimney being built for a wood stove. You can see the plumb bob hanging where the thimble will go. The hole is where the horizontal section of fireclay liner (called the thimble) hooks into the vertical flue. The stove pipe will fit inside this thimble and to the inside edge of the chimney flue. Codes call for at least 8 inches of fireclay liner below the entry point of the thimble to ensure a safe distance between the intense heat and the vulnerable masonry.

The chimney in FIG. 9-7 was built against a temporary plywood backing, which is held away from the wall by temporary 2-x-4 nailers. When the chimney was done, the studs were knocked out and the plywood was removed. The free-standing stone chimney was then 2 inches from the frame wall of the house. The space will ensure that the stone will never heat the wooden wall enough to ignite.

The only place the fire codes permit a masonry chimney to touch wood is the exterior siding, which

Fig. 9-7. A plumb bob is hung from above to set the liners plumb.

can abut the masonry. That code is being challenged, and soon, all masonry chimneys might need to be isolated from combustibles by at least 2 inches, even on the outside of a house. Where the chimney goes through a ceiling or roof, the exterior of the masonry must remain at least 2 inches from the framing or other combustibles. Where the masonry penetrates the roofing, there must be 2 inches of air space around the masonry.

The gap between the masonry and roofing must be bridged with flashing metal. The flashing is an extension of the shingles or other roofing, which climbs the vertical surface of the chimney and tucks into a mortar joint on the chimney, near the roof level. As for the shingles, lay the lowest flashing strip or shingle first. Each shingle of metal must extend onto the horizontal part of the roof enough to have a place for you to nail into the decking. These strips of metal are the only place the building is actually in contact with the chimney. They are necessary to keep the rain from entering around the chimney opening.

Cheap flashing metal is often covered over with a facade shingle. Caulk nails holding the facade shingles and the flashing to the roof deck.

FIGURE 9-8 shows a stone chimney and the flashing that connects it to the raised-rib roofing. On the uphill side of the chimney, a cripple has been built

Fig. 9-8. A cripple is used on the high side of the roof to repel water from the chimney.

to divert water away from the high-side of the chimney. The sides of the chimney are flashed in steps, starting at the bottom, a process called *step flashing.* With rubble stone, you need to make a point of leaving some nice horizontal joints in the area where the flashing will tie in. They aren't necessary with brick or block chimneys because you can't avoid the level horizontal joints.

Clean-out Openings

When you build a chimney, you should include a provision for removing the accumulated chimney sweepings. The clean-out door must be able to be closed and be made of iron or steel. Place it at the level of the bottom flue liner, beyond the thimble openings. It should be accessible without being very noticeable.

FIGURE 9-9 shows a clean-out arrangement on an inside chimney. The metal door is mortared flush to the masonry. The prefabricated doors and frame have flanges to grip the mortar. The door is attached to a horizontal section of flue liner, which comes into the side of the first vertical section. This arrangement will require a long-handled shovel to remove the sweepings, but the wood stove that will go in front of the stone wall would block the access if the clean-out door had been on the front face of the chimney wall.

The horizontal run for the clean-out opening doesn't need to be done with flue liner. You could maintain the opening by the positioning of the stones. In this case, the liner was handy to maintain that space and to prevent mortar from above from falling in. It also provides a smooth surface to slide the little shovel on.

FIGURE 9-10 shows another chimney, with the clean-out door in a more typical place. Notice the block foundation under the stone. This block will be below the finish grade, and it extends down to the footing. This was done to save time on stonework. Notice the clean-out door is spanned by a long stone. There is no real weight on top of that door. The opening extends through 8 inches of stonemasonry, then through a few inches of block, and through a hole in the side of the bottom flue liner.

FIREPLACES

If you have definitely decided you want a fireplace, and not a wood stove, furnace, or other heater, there are still a few ways you can use to get around actually building a masonry fireplace. A rubble-stone fireplace isn't a beginner's job, although there are certainly people who could build one with no experience. For a novice, it is much easier to use a prefabricated fireplace. Some models require no clearance from combustibles and need no masonry cover. If you mostly want to view a fire and aren't too concerned with heating efficiency or traditional aesthetics, a prefabricated fireplace could be for you.

Steel-lined Fireplaces

If you want more of the traditional feel of a masonry fireplace, but are shy to tackle the skilled work, there is another option: a steel-lined fireplace. These units mimic a masonry fireplace and have all the same parts, made of steel. They are designed to be covered with masonry, and use a masonry flue (FIG. 9-11). In some ways, they are better than a regular fireplace because of the duct work in the layers of metal, which pull cool air from floor level and put out warm air at the outlet. These air passages can be assisted with a fan.

The only visual difference between a steel or masonry fireplace is that you see blackened steel in the back of one firebox, and blackened firebrick in the other. Don't expect a steel fireplace to last as long as the masonry, however.

Masonry Fireplaces

New and stricter fire regulations and building codes are making it difficult to build a standard masonry fireplace. Upcoming pollution regulations will make it tougher yet. An irony of a masonry fireplace is that it is basically a very primitive technology being forced to fit into an overall more refined technology. It would be easier to start from scratch with a totally new design than to try to add every conceivable gizmo to a conventional fireplace in order to get more heat, less danger, and less pollution.

The Rumford design, with its tall, narrow, and shallow firebox is fairly efficient at radiating heat directly into a room. The design does not meet most minimum depth codes, however. Baffled fireplaces, like the Russian type, are made with walls one brick thick. This is too thin for many building codes, but the fireplace wouldn't work well with thicker walls.

A chimney that will have a masonry fireplace will need a larger footing than one that merely has a wood stove. A minimum of 6 inches larger than the fireplace on all sides is standard. FIGURE 9-12 shows a cutaway view of a standard masonry chimney, with the different parts labeled.

Old fireplaces were often built with massive wooden lintels. The *lintel* is the beam that carries the masonry above the opening, and a wooden one has been deemed a fire hazard. I think in the old fire-

Fig. 9-9. A clean-out door and opening are used at the end of a wall in this example.

places, the wooden beams were okay because the openings were so large (some of the fireplaces had seats inside) that the wood was always pretty far from the flames. Codes now say that a wooden mantel must clear the height of the opening by at least 12 inches. When wood stoves are put in front of fireplace openings, you need 36 inches of clearance between stove and wooden mantel.

Masonry has so much mass, it rarely gets hot on its surface. Metal stoves get much hotter on the surface, and need wider hearths than fireplaces. Fireplace hearths should extend 8 inches to the sides and 20 inches in front of the opening, to be safe. The hearth must have no wood in direct contact on the underside, so if you use forms to build the concrete hearth, you must remove them when you are finished.

The idea of all these fire codes is to keep flammable materials at a proper distance from the hot parts of your wood-burning scenario. All the codes are folly if you choose to put a thick, highly flammable carpet in front of the hearth and it catches fire from a long spark and an unattended fire. Many people leave their kindling and paper closer to the fire than codes would allow, but that sort of thing can't be regulated.

PARTS OF A FIREPLACE

FIGURE 9-13 shows another cutaway view of a fire-

place with the distances of different parts indicated by the line with a letter. The accompanying Table (TABLE 9-3) shows the dimensions for the various parts for different sizes of fireplaces. These figures are the conventional. Fireplaces and chimneys have been made of every possible shape and size.

The Base of the Fireplace

The *base* of the fireplace includes the foundation or footing and the hearth support. In the area between is the ash pit, the clean-out door, and an outdoor air intake. A separate air intake for any kind of wood heater is a good idea because a fire needs air. If the air isn't provided for at the scene of the fire, it will be drawn from cracks around the doors and windows, and the cold draft is felt. If the fire is able to pull its fresh cold air from underneath and behind the firebox, you won't feel the cold.

Sometimes, a fair amount of wasted space must occur to have a fireplace opening well above grade.

Fig. 9-10. This chimney has an outside clean-out door.

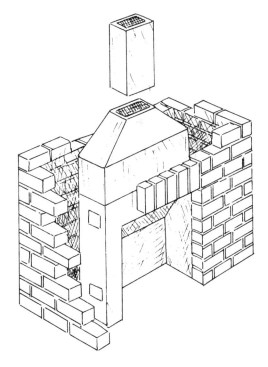

Fig. 9-11. Steel-lined fireplaces can be made to look like masonry.

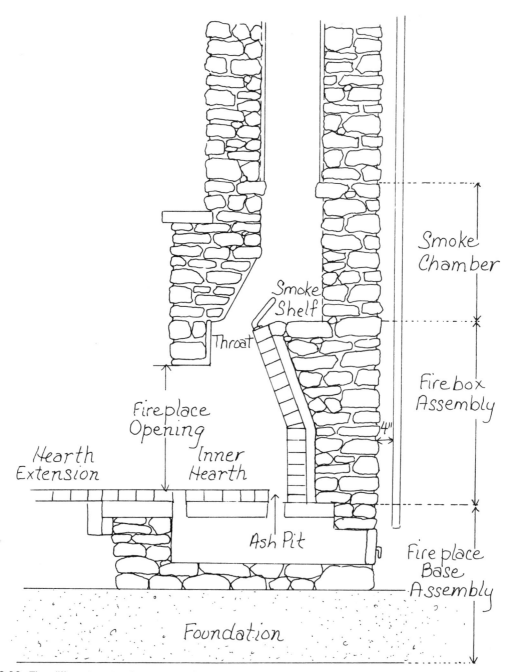

Fig. 9-12. *The different parts of a masonry fireplace.*

A heavy masonry structure must have support clear down to the subsoil, even if it is only needed on the second floor. I have seen brick chimneys supported by wooden posts at second-floor level in old houses.

It is hard to recommend, and probably not able to pass codes.

A fireplace in the basement of a house might not have enough available altitude to offer a drop-down

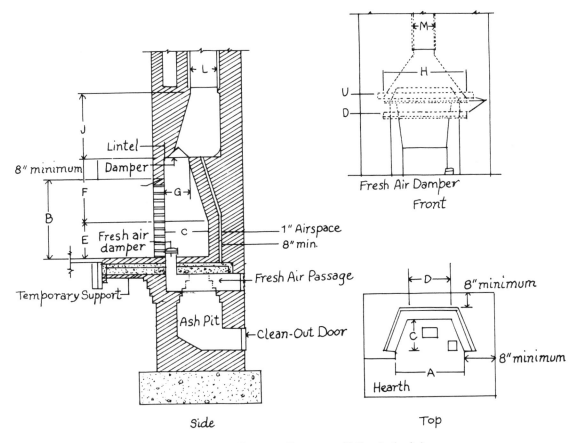

Fig. 9-13. Conventional fireplace dimensions (Courtesy Tennessee Valley Authority).

Table 9-3. Conventional Fireplace Dimensions (in inches).

CONVENTIONAL FIREPLACE DIMENSIONS
(in inches)

GENERAL DIMENSIONS									Flue Sizes		Lintels	
A	B	C	D	E	F	G	H	J	L	M	Smoke Chamber (U)	Profile (D)
24	24	16	11	14	18	9	32	19	8 × 12		3 1/2 × 3 1/2 × 1/4 × 38	3 × 3 × 3/16 × 36
26	24	16	13	14	18	9	34	21	8 × 12		3 1/2 × 3 1/2 × 1/4 × 40	3 × 3 × 3/16 × 36
28	24	16	15	14	18	9	36	21	8 × 12		3 1/2 × 3 1/2 × 1/4 × 42	3 × 3 × 3/16 × 36
30	29	16	17	14	23	9	38	24	12 × 12		3 1/2 × 3 1/2 × 1/4 × 44	3 × 3 × 3/16 × 36
32	29	16	19	14	23	9	40	24	12 × 12		3 1/2 × 3 1/2 × 1/4 × 46	3 × 3 × 3/16 × 42
36	29	16	23	14	23	9	44	27	12 × 12		3 1/2 × 3 1/2 × 1/4 × 50	3 × 3 × 3/16 × 42
40	29	16	27	14	23	9	48	29	12 × 16		3 1/2 × 3 1/2 × 1/4 × 54	3 × 3 × 3/16 × 48
42	32	16	29	14	26	9	50	32	16 × 16		3 1/2 × 3 1/2 × 1/4 × 56	3 1/2 × 3 × 1/4 × 48
48	32	18	33	14	26	9	56	37	16 × 16		3 1/2 × 3 1/2 × 1/4 × 62	3 1/2 × 3 × 1/4 × 54

ash pit and clean out. Ashes would need to be emptied from the front, or they could fall into a removable container that sits in a recess in the inner hearth.

The hearth, according to the new codes, may not be supported directly underneath by a wooden floor. It can rest on a masonry floor, or it can be a cantilevered, reinforced slab, built with wooden forms that are later removed.

The Firebox Assembly

The *firebox assembly* includes the fireplace opening, the combustion chamber, the throat, and the smoke shelf. If you ever think you might want an insert for your fireplace—and thousands do—you should size your opening according to standard sizes for which inserts are made (TABLE 9-3).

The opening of a fireplace needs something to support the masonry above. The most common approaches to crossing that space are steel lintels, masonry lintels, and masonry arches. Masonry lintels are rare in fireplaces. The masonry usually isn't very strong in shear strength, and in a tall chimney, quite a lot of weight might need to be carried by the lintel.

FIGURE 9-14 shows a fireplace under construction with a big stone lintel. With its height, this stone could carry a lot of weight. In this case it doesn't need to because I have built a relief arch above it. This arch will spread the weight of the tall chimney to the sides of the opening.

If you use an arch as a lintel, you must provide adequate mass at either end of it to resist the load carried by the arch. This isn't a problem with fireplaces that have a whole wall of masonry. The most common, and the easiest way to get across the opening is with angle iron. All but the edge of the angle iron is hidden from view, yet it can nicely carry a brick facade. It is harder to fit rubble stone on an angle iron ledge.

At either end of a metal lintel, there should be a little scrap of mineral wool to ensure a little space

Fig. 9-14. This fireplace has a large stone lintel, and a relief arch to carry the weight over the opening.

for the expansion of the metal lintel, which will be greater, and happen faster than the expansion of the masonry. When the heat is on, it will crack the brittle masonry if there is no room for the iron's expansion.

The Hearth

The *inner hearth* of a fireplace is actually the floor of the combustion chamber, and is made of firebrick with very thin joints of fireclay mortar. The *hearth extension* is the part that comes out beyond the face of the fireplace. Minimum dimensions for the hearth distance from the fireplace wall are 16 inches for a small fireplace, and 20 inches for a larger one. Why cramp a fire? A big hearth is both safer and more aesthetically pleasing. Type N mortar is adequate for the hearth extension.

The Combustion Chamber

The *combustion chamber* is where you put the wood to burn. The shape and angle of the firebrick in refractory cement mortar is the most demanding part of the job. The hard bricks need to be cut, some of them in two directions, and the joints must be kept very thin—1/8 inch or less. These firebricks for the back of the combustion chamber are normally laid using an angled template of plywood to temporarily lean them against something of the proper angle.

The angles and proportions of the back and sides (coving) of the combustion chamber are crucial to the proper functioning of the heater. An entire book could be written about how to build a masonry fireplace. Because this book is for the novice rubble mason, I will go into no more detail here about laying the firebrick.

Throat

The *throat* of a masonry fireplace is actually part of the firebox. It consists of a slot opening above the main firebox, where the flames and smoke pass into the smoke chamber. The throat design will have an important effect on the way the draft works. The throat is where the damper goes. The *damper* is usually a heavy metal item, laid in fireclay mortar, with enough space left for the expansion of the metal.

You should install the damper so it can be replaced, since it will wear out sooner than the rest of the fireplace. I have seen masonry fireplaces that offered no way to replace the damper. Actually, a damper is more effective on the top of the chimney, operated by a chain that comes down the flue. A damper in that position will allow more heat to be held in the chimney, and it will be much easier to replace later. Chimney-top dampers are used less often than throat dampers, but you should consider them.

Smoke Shelf

The *smoke shelf* is the last part of the combustion box assembly. A smoke shelf's purpose is to direct stray downdrafts back up the chimney so that smoke doesn't enter the room. The smoke shelf extends the full width of the throat, causing an obstruction to the vertical flow of smoke.

Smoke Chamber

The *smoke chamber* is not part of the firebox assembly, but is rather a funnel-shaped area leading to the chimney proper. The shape and angle of this chamber facilitates the flow of exhaust up the chimney. The back of this chamber is vertical and the side walls slope toward the center. The front wall is sloped enough to support the edge of the flue liners. The first flue liner should be supported on all sides, without any of the support extending into the flue space.

It is important to parge smooth the interior surfaces of the smoke chamber, both to reduce the amount of drag on the escaping gasses, and to lessen the likelihood of deposits forming. Use a broad trowel and refractory cement mortar for the parging. Regular mortar cannot hold up to the intense heat and temperature changes in this zone.

Parging is a detail often left unfinished because of the awkwardness of getting in the fireplace, but it is important for the proper functioning of the fireplace. A fireplace and chimney is just a long air tube, and it should be as smooth as possible inside, from top to bottom.

CODES FOR MASONRY FIREPLACES AND CHIMNEYS

A lot of the construction of a fireplace/chimney is fairly obvious masonry procedure. Near vertical alignment is important; a solid footing below frost line is needed; use of solid materials is essential; and so on. The less obvious aspects have to do with the fire containment details. Clean-out doors, for example, should have ferrous metal doors that will close tightly. The opening for the clean out should be at least 8 inches below the nearest thimble entrance. Refractory cement mortar is a must where high temperature swings will occur.

Expansion Space

The air space around tile liners is essential to allow for expansion in the liners. This is a detail very often overlooked in chimney construction, as is the use of fireclay mortar. With bricks, it is easy to maintain this air space around liners. With rubble stone it is more of an effort. Fiberglass insulation can be helpful if you lay it against the liner before you put the stone in. The mineral wool will hold back the mortar between stones and maintain the small space required.

This space has nothing to do with air flow; it is simply to provide room for expansion of the liners. Without the space, the liners will crack, and they are nearly impossible to replace once the chimney is in.

Corbelling

Corbelling is a way to leave vertical alignment with masonry structures without creating a new footing directly below the new masonry. It is done in small steps (FIG. 9-15). New building codes call for no more than a 1-inch extension on each course of corbelled

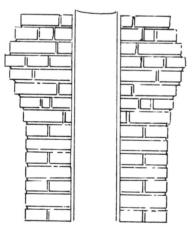

Fig. 9-15. You can't corbel further laterally than the thickness of the chimney walls.

bricks, with the total amount of corbelling not to exceed the thickness of the chimney walls.

A corbelled chimney needs a thick wall to begin with to support the renegade bricks. If the walls of the chimney are 12 inches thick, generally you could get away with 12 inches of corbelling, but it would require the altitude of 12 courses of brick to get that 12-inch extension, protruding 1 inch per course.

When it comes to rubble stone all the codes get a bit vaguer because the material itself defies definition. In the case of corbelling, I would think the same holds true for stone, but since the courses are not of uniform altitude, you could shoot for the same basic proportions of a brick corbel, as seen in FIG. 9-15. When corbelling, don't strand small pieces on the edge of the wall; keep them safe inside the wall and let a full-size unit do the actual overhanging.

The codes imply that it takes 12 inches of mountain stone to achieve the stability of 4 inches of brick. Rubble stone should not be thought of as an exact item, even in its nonexactness. Really good stones wouldn't need 12 inches to make a stable chimney wall, and really bad stones wouldn't achieve stability in 12 inches. If your rubble stone chimney walls need to be 12 inches thick (minimum) and you are using a 12-inch flue liner with the air space around the liner, you are talking about a minimum thickness of 36 inches for a stone chimney. An 8-inch flue liner could make it slightly smaller.

Corbelling in chimneys is done for several reasons. At the top of the chimney, the last few courses are often corbelled to form a drip edge, to minimize the amount of rain that actually runs down the chimney face. Another reason for corbelling a chimney is to create a vertical stagger to get around part of the framing of the house. I don't go for that reason too much. Wood should be adjusted to accommodate sound masonry, not the other way around.

If you must create a nonplumb chimney, any sudden changes in the direction of the flue effectively decrease the overall size of the flue to the size it becomes in vertical profile at the point of nonverticality. In other words, you need a bigger flue if you are going to make it twist and turn. Codes call for these directional changes to be made at least 6 inches from any framing wood because the bends are prone to creosote accumulation, and you don't want any of the house wood near a creosote fire.

Height of the Chimney

The height requirements are described in FIG. 9-6. Generally, a chimney needs to be at least 3 feet high where it leaves the building, and 2 feet higher than any obstruction (like the roof) that is within 10 feet. In some areas, this will be inadequate to facilitate a healthy draft. If your house is at the bottom of a valley and surrounded with tall evergreen trees, you will need a taller chimney than the minimum set by the codes.

Multiple Flues

Chimneys with more than one heating appliance must have separate flues for each. The flues need to be separated by at least 4 inches, including the air space. Multiple-flue chimneys provide a more efficient use of materials than separate complete chimneys for each heater. Two side walls can be eliminated in the shared chimney. Multiple-flue/chimneys also give rise to the wide, nonsquare shape that seems to be more pleasing than a square chimney.

Clearances

Fire-stops between chimneys and combustible materials such as floor or roof rafters should be made of galvanized steel of 26 gauge or thicker. The exterior of an approved chimney must have at least 2 inches of gap from the combustible framing of the house. The exception to this rule is on the exterior of a masonry chimney, where the siding is allowed to touch the outer masonry. This code is being challenged and might be subject to revision, requiring clearance even on the siding.

Thimbles

Where stove pipe is to enter a chimney, you should use a fireclay thimble (usually 8 inches in diameter). Fill the space between thimble and stove pipe with a noncombustible ceramic wool. Don't fill this area with solid mortar, because mortar will allow the thimble no expansion room. Thimbles, and the stove pipe that goes through it, should extend into the vertical flue liner, but not further than its inside surface. A thimble must be surrounded by at least 8 inches of masonry.

If you are not using a thimble, you can use an opening at least 18 × 18 inches, with a piece of sheet metal having a hole in the center for the stove pipe. Attach the sheet metal to the framing to close off the opening.

Common Sense

There are many aspects to building and operating a fireplace and chimney that are not covered by building and fire codes. So far, there is no safe residential way to have a chimney fire. Admittedly, certain designs have a better chance of surviving intact, and going by the book will lessen your chances of burning your house down, but none are fail-safe.

If the codes want more safety than they are presently bringing, all fire burners would have to operate at above 1000°F or be equipped with a catalytic combustor. Currently, people are allowed to burn a dirtier fire in their homes than would be allowed for industry or power plants.

TROUBLESHOOTING FOR CHIMNEYS AND FIREPLACES

Frequently, people use a fireplace because the old house they live in came with one. Often, these old fireplaces don't work quite right. Following is a summary of typical problems and solutions regarding masonry heaters.

House and Chimney Separation

If the house and chimney seem to be separating, either the chimney is sagging on its footing, or the house is moving away from a stable chimney. Sighting along the edge of a level plumb will tell you if it is the house or the chimney that's leaning, assuming either were built plumb. There is a good chance that the chimney footing is either nonexistent, not deep enough, not thick enough, not broad enough, or being undermined by erosion.

An unoccupied 100-year-old barn near my house has been wrecked because the gutters were still working but the downspout had become unconnected. Because the area of barn roof was so large and because it is a very tall barn, the water that would spill off the end of that gutter run during a hard rain would come down like a waterfall. The waterfall began to dig away at the pier footing until the pier slid into the hole. That allowed a major beam to fail and a floor to break apart. If the water had fallen off the entire end of the roof, the waterfall would never get intense enough to do much harm, but when you gather all that water into one spout, it can do some fast damage in a very few rains.

If you find that erosion of some sort is the problem, you need to correct the water flow before you do anything else. Some people will strap a leaning chimney to the house. That makes me nervous. If the chimney really wants to fall, it would be better if it didn't bring the house down with it. The drainage needs to be fixed and the footing enlarged.

Keeping 20 tons of chimney in place while you try to fix the footing is no job for an amateur. It's not really great for a professional, either. If you are willing to merely keep the chimney from sagging any further, you could pour new concrete in an excavation around the old footing, hoping that it will fill the hollows under the old footings and tie it into a greater mass. The new concrete could come up higher and over the old footing.

Charred Lintel

Another common problem with an old fireplace is a charred wooden lintel. You must remove the wood and replace it with masonry or metal. You need to keep combustibles at least 18 inches from wood stoves and stove pipe, and that includes at the floor. You also must provide temporary support for the masonry above the outdated wooden lintel as it is replaced.

Fireplaces of historical significance with wooden lintels shouldn't be changed. If you use one, pay attention to the size of the fire and how hot the wood feels to the touch. If wood is too hot to touch, tone down the fire quickly.

Missing Items

People run into trouble trying to fit an insert or a wood stove into their fireplace. You can make an existing fireplace smaller, but you can't make one larger. You could find a metal worker that could make an insert for an odd-sized opening, but don't try to

rearrange the firebricks. If the firebricks are damaged, a good mason can replace them. If only the mortar is damaged, anyone can replace it with some fireclay mortar.

Occasionally, you find a fireplace that has no smoke shelf. If it smokes, there's not much you can do about it. Try a wood stove with a pipe that goes beyond the smoke shelf area, and fill the space with ceramic wool. A good rain cap can prevent downdrafts and might be worth a try before making any big expenditure.

Flue Problems

Sometimes there is a gap between the flue liner at the bottom and the wall of the smoke chamber. This gap will allow smoke into the room. You can patch this area with refractory cement mortar, or stuff it with ceramic wool.

Other flue problems might be that the flue is too small to handle the smoke. More altitude might help, but there's nothing you can do about the size of the opening in the chimney, except make it smaller.

When a flue liner is cracked and busted, it is very hard to replace it. Sometimes with a brick chimney, you can take part of the brick wall out to get at the old liner and slip a new one in. There are businesses that reline old chimneys with specialized equipment.

Many old chimneys have no flue liner, and it is not easy to put one in after the fact. It is easier to put a stainless steel pipe in an old chimney. Flexible stainless-steel pipe has been developed for the purpose of relining old chimneys.

None of this work is for the weekend tinkerer. In general, an archaic masonry fireplace and chimney, especially if they are in need of repair, are best used if retrofitted for a wood stove. The chimney should easily accept stove pipe, which will relieve the masonry of the thermal strain.

Crumbling Mortar

Fireplaces and chimneys that have surface mortar crumbling out can be repointed by the casual builder. Chapter 12 gives the details of the repointing procedure. If you do chimney repair on a roof scaffolding (FIG. 9-16), it is important to spread the weight of the scaffold on a strong wooden wedge that levels out the angle of the roof. Nail a strip of sheet

Fig. 9-16. Scaffolding on roofs must have a very broad seat to spread the weight and bring the scaffold up level.

metal to the wooden "jack," and nail that strip under a shingle to keep the stand from sliding down the roof.

Scaffolding on a roof deck should be loaded very scantily. The roof isn't made to hole a ton of weight on a few square feet. Don't lean a tall ladder against a stone (or brick) chimney having loose masonry units. The tops of any masonry walls not covered with a large cap on a building are vulnerable. The lower part of the chimney is more stable because of the weight above, so you could lean a ladder against a chimney if it hit the chimney several feet from the top.

A single unit of scaffold will fit around a small chimney, and you can set up a rectangle of beams and planks to service all sides of the chimney. As seen in FIG. 9-16, the boards are actually not on the scaffold, but are on beams that cross the scaffold.

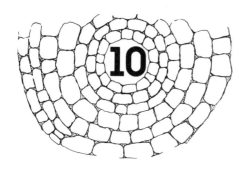

10

STONE STEPS

STEPS made of stone can be beautiful, practical, inexpensive, and within the grasp of the weekend handyman. A shovel and a wheelbarrow might be the only tools you will need. A small set of steps (two or three) could be done in one day.

Stone stairs are not something to rush through. It will take a lot of adjusting to get rubble stone to feel right to your feet when you step up or down. If you don't get it right, you won't want to use the steps. Stairs that wiggle give an insecure feeling. Steps that continually vary in tread height and depth are hard to negotiate in the dark. Stepping stones with depressions on top or steps that slope inward will collect water and ice, so they should be avoided.

There are two basic types of steps: ones that more or less follow the slope of the ground, giving the climber a level place to step, and ones that reach up into space, like a set of steps from the ground to a porch. There are many different approaches to building steps; for instance, with and without mortar. There are many exotic forms of stone steps, such as the large slabs that were left protruding in strategic positions in tall walls of Ancient Rome. They could be used as steps on the side of the wall. The weight of the stone wall held the stones in place, although they had no railing. The famous Lost City of the Incas at Machu Picchu, which in my opinion has the best stonemasonry on the planet, has stone steps up the mountain that are chiseled out of the granite. They are probably the best stone steps on earth, but few are in a position to emulate the method used.

STONE STEPS
THAT FOLLOW A HILL

The type and quality of the steps you build will be determined largely by what you need them for. FIGURE

10-1 shows a very steep yard. This lot already has concrete steps with a railing as a main access. The woman who lives there wanted to do something to prevent erosion on the hill, so she chose to plant cover crops. In some places, the slope required low, dry-laid walls to terrace the ground. The woman bravely took up the shovel and gathered stones in the backseat of a compact sedan.

A cover crop cannot be effective if it never has a chance to take hold, and it won't take hold if it gets walked on. A crude set of stone steps was in order to get at the plants and to get up and down the hill without more erosion. The steps (FIG. 10-2) are the access to maintain the plants and shrubs on the bank. With no experience in construction, and only a shovel and wheelbarrow for tools, this woman transformed a steep, barren slope into a beautiful and stable environment. The small stone wall in the upper right of FIG. 10-2 enables the side walls of a partially underground garage to be covered from sight. The stepping stones, although of rough surface and at odd intervals, are an extremely practical solution for the problem presented here. The project cost nothing. For occasional access, the rough stepping stones are great, and they blend more readily into a nature walk sort of area.

Rubble stone is probably not the best choice of materials for a heavily used major access. It is difficult to find smooth large stone of the proper dimension. Single large stones are the best for a walkway laid in dirt without mortar. If the stone outweighs the person stepping on it by quite a lot, there will be less chance of it tipping up from weight on the leading edge. With smaller stones in steps, this is always a problem, even when they are "glued" down with mortar.

In FIG. 10-2, the seventh and eight stones from

Fig. 10-1. A steep yard is helped with some stone walls and steps.

the bottom might kick up because there is a hollow underneath them. These spaces could be filled with stones and dirt. A stone is more stable, however, if it is partially resting on the back of the stone below. Very heavy and deep stones could handle a certain amount of this overhanging without tipping down.

Stepping stones are certainly the fastest way to get a crude stairway up a hill. If you want to take the next step in refinement—if you want to be able to carry a weight up the steps without looking where you are stepping—you will need to plan out the steps so that each successive step is partially sitting on the one below it. This type of construction must start at the bottom, because the stones are stacked. Stepping stones aren't stacked, needn't be started at the bottom, and probably don't qualify as masonry. For the more refined, yet crude steps, a larger area than the tread must be excavated for each step because part of the stone will be overlapped by the next higher step. The greater the percentage of each step that gets overlapped, the more stable the steps.

Rise and Run

Suppose you have a house on a steep hill and there is a stream downhill that you frequently walk to. Walking directly to that special spot has created a worn path where no grass can grow. When it rains, the path gets muddy, slippery, and washes away. With luck, the stream is strewn with large, flat stones, or

Fig. 10-2. The most basic set of steps adds a useful accent to the steep yard.

maybe the hillside has many boulders. If so, a stone stairway could solve the problem for the cost of the extra food you'll eat while putting the rocks together. To build a good stairs, it helps to calculate the size and number of steps you'll need. To do that, you need to be familiar with rise and run.

The *rise* of a set of stairs is the total altitude gained from the bottom of the first step to the top of the last step. In a house with floors 8 feet apart, the rise of a set of stairs would be 8 feet. The *run* of a flight of stairs is the total distance covered horizontally. If a set of steps has an 8-inch-high step, then an 8-inch-wide tread, the rise and the run would be the same, and the steps would travel at 45 degrees. A straight up-and-down ladder is all rise and no run. A level floor is all run and no rise.

To calculate the number and dimensions of uniform stairs, you need to know the rise and the run of the particular situation. If you were building a wooden staircase for a house, you can measure the rise and run fairly easily. The rise is the distance between finish floors, and you can find the run by marking a spot on the first floor where the stairs will start, marking a spot on the second floor where they will stop, hanging a plumb bob from that point, making a mark on the floor, and measuring the distance between those points. That is the horizontal distance covered while traversing the stairs.

Number of Steps Needed

To find the number of steps you need, divide the rise by the desired step height. Suppose you had 8 feet between floors and you wanted stairs with an 8-inch step up. 8 feet = 96 inches; 96 ÷ 8 = 12. It would take twelve steps. Now suppose you had an attic 8 feet higher, and you wanted steps to it that were 12 inches high instead of 8, to save work. 96 ÷ 12 = 8. You would need only eight of the taller steps.

Tread Depth of Each Step

To find out how deep your steps must be where you put your foot, divide the run by the number of steps. Suppose you only had 8 feet of space in which to fit a flight of stairs that rose 8 feet, with twelve 8-inch steps. The run is 96 inches divided by the number of steps, which is 12. Eight inches is the tread depth.

Sometimes the run of a flight of steps is not predetermined. In this case, the run can be adjusted to accommodate a certain tread depth. For example, suppose you need to get from ground level up to a deck that is 32 inches off the ground. If you want an 8-inch step height, divide the rise (32 inches) by the desired step height (8 inches) to get the number of steps needed, which is four. To get the tread depth, divide the run by the number of steps. Because the run is undetermined, you can determine it by choosing a comfortable tread depth. If you like 12 inches to step on, multiply that tread depth by the number of steps (four) to get the run. The run is 48 inches.

Suppose the rise in a similar scenario were 36 inches. The number of 8-inch steps needed to gain 36 inches of rise would be 36 + 8 = 4.5, which is a problem. The half step would be a hassle to negotiate. It is best to have steps of equal dimensions. In the case of the 36-inch rise, you should have either four or five steps, not four and a half. If you decided on four (less work), you would divide the rise by the number of steps to get the individual step height. 36 ÷ 4 = 9. A 9-inch step would come out even.

Generally, to build steps, you must adjust the variable factors, or the undetermined factors, to come out with a uniform dimension step. Normally, the rise is predetermined. You want to get from one determined altitude to another. The other factors are flexible to a point, although individual step heights probably won't be more than 12 inches nor less than 6 inches. The tread depth is a bit more flexible, although if it gets much narrower than your foot, the stairs will be hard to walk on.

Sometimes the step height is also a fixed dimension. If you happen to have a bunch of 10-inch-thick cut granite slabs, you will want to adjust the other factors to accommodate these rocks, rather than trying to add or subtract from the stone to make it fit in a situation with an arbitrary rise and run. Stones can be made slightly taller with a bed of mortar or a smaller stone underneath. They can be made effectively shorter by burying them partly in dirt, although this won't work if the stones are to overlap. In other words, you won't be able to use stones thicker than 10 or 12 inches, except at the bottom, where you could bury part of an extra-thick stone.

In a short run of steps up to a deck, the top step might not need to go all the way up. The deck itself could provide the top step. This would change all the calculations. The rise would only be counted to the top of the last step, which would be one step short of the deck. If the deck were 30 inches above grade and your desired step height were 10 inches, it would take three steps to get to the deck level, but only two steps to get one step short of the deck. If your run were 30 inches also, it figures that each tread depth must be 10 inches to bring you to the deck. Because you are using the landing as the top step, the real run in this situation is only 20 inches, and two steps of 10-inch tread depth will get you there.

FIGURE 10-3 shows a rustic cabin with stone steps to a deck (lower left corner). There are only two stones, but there are three steps, because the deck is acting as one. So to find the height each stone must go, you would divide the height to the deck by three, not two. If you use a deck or landing as a step, then figure your rise one step short.

The cabin in FIG. 10-3 is built deep in the woods, of local logs, stones, and boards. Only the glass, shingles, and nails were brought in from afar. The building sits on logs, which are on stone piers, which have had underpinnings of stone added. The steps to the deck needed no excavation. The bottom step is so big and flat, it doesn't matter if it moves a little in a frost heave. It won't break up like a smaller step of mortared stones would do. Part of the bottom stone, where it is too thick, is dug in the dirt. The second step is part of a stone wall that was left protruding. The top step is the deck.

If you create a step by leaving a stone sticking out of a wall, you must be sure to have that stone fit deeply in the wall, and it must have enough weight on top to counter the weight of people standing on the overhanging ledge. The stone used this way must be thick enough and of good shear strength so it does not crack from stress. Many people assume that all stone is very strong, when in fact it is often much weaker than wood, inch for inch, as a beam. If I were in a karate exhibition, I would rather try to break a 1-inch board of stone than a 1-inch board of oak if

Fig. 10-3. This cabin has a stone foundation and stone steps.

they were both spanning the same distance. You'll do fewer dumb things with rocks if you think of them more as glass than as steel.

As in FIG. 10-3, there are many situations where you only need two steps to get from the ground to a building. When this is the case, you might as well customize the situation to fit your stones. In two steps, a person hasn't had a chance to develop a pattern, so it won't matter as much if one stone has a wider tread than the other. Also in a two-step situation, the steps can be independent units, as seen in FIG. 10-4.

The steps in FIG. 10-4 were something of an afterthought, so I missed the opportunity to make the second step part of the foundation wall, below the door. The owner of this project had that stupendous slab of hard sandstone for a step, and it also makes a nice landing. The bottom stone sits on the ground, and is a reasonable height for a step. The platform stone was set at twice that height to give even steps. The third step, which is up into the house, is a shorter one, but the awkwardness is broken up by the door itself. You take two even steps, stop, open the door, and take a shorter step. From the inside going out, no pattern can develop, so the stairs are comfortable even though the height, width, and tread depth vary to accommodate special rocks. A long flight of individual stones would be tedious.

Fig. 10-4. These steps are independent of the house and each other.

If you build independent steps as shown in FIG. 10-4, each stone should be big enough to withstand the tipping force of weight on the edge. The bottom block of granite in FIG. 10-3 weighs almost 300 pounds. It is not going to move when you jump on it, and it is not going to break in two if the frost pushes it up. The second step is mortared, and it needs a footing so the mortar won't crack from motion. The foot coming out of the stonework is ceramic.

For the most part, short flights of rubble stone steps don't need to be very exact, and won't be very exact even if you want them to be. People sense the roughness of stone and behave accordingly. Inside a house, you might be able to run up a flight of stairs blindfolded, but on a set of stone stairs, you tend to watch your step.

It is quite possible to build stone steps without regard to rise and run calculations. Instead, you pay attention to the particulars of the slope and the rocks, and build one step at a time. On a long walkway down a steep hill that becomes more gradual, you might want a flight of steep steps, followed by a horizontal landing, followed by a more gradually inclining set of stairs. This type of stairs could have charm and grace in a rustic setting. Rather than worrying about the number of steps coming out even on a long set of stone steps, you could just start building the steps to a height you find comfortable, like 8 inches. When you get to the top, you can readjust the dirt a little if need be to avoid using a step of a fractional height.

If anything is to be uniform in a set of steps, it should be the height. The tread depth of the steps can be changed, preferably after a series of same ones, to accommodate changes in the ground slope. The height of each step can remain constant through other changes (FIG. 10-5). If you want rubble stones steps to carry you up a hill of varying slope, decide on a step height. Eight inches is comfortable, but for crude steps you might want to go for 10 to 12 inches because it is less work. Go for the highest comfortable step.

Suppose you decide on 10-inch steps. If your slope is roughly 45 degrees, the tread depth of each step would be the same as the rise: 10 inches. Slopes less steep than 45 degrees imply steps with a tread depth deeper than the height of step. Slopes steeper than 45 degrees imply steps with a tread depth narrower than the height of the step. Steps can wind up

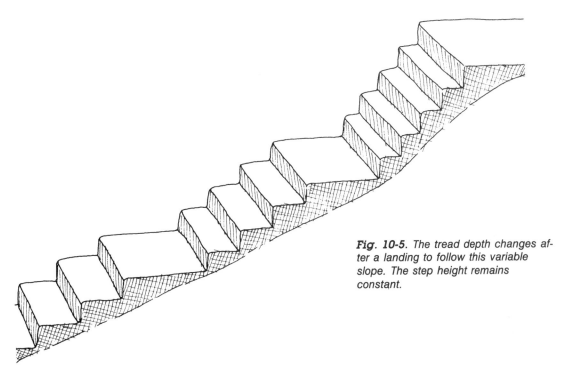

Fig. 10-5. *The tread depth changes after a landing to follow this variable slope. The step height remains constant.*

a hill to make the slope less so the tread depth doesn't need to get too small to stand on.

Using a protractor and a level, you can get a pretty good idea what slope the land is where you want steps. Holding the base of the angle finder on the level, sight through the protractor to see which angle most closely aligns with the land. It isn't essential to good steps, but if you can calculate the rise, you will know how many steps you'll need and the project will be planned better.

Finding the Rise in a Crooked Situation

Suppose you have a lake-front house on a hill, and you want a flight of steps from the porch of the house to a dock on the lake. The easiest way to find out how much higher the porch is than the dock is to use a transit. If you can't get one, there are some cheaper tricks you could try.

One trick is to put a level on the deck, adjusting it to aim at the dock and to read level. Meanwhile, on the dock, your helper holds a very long, straight pole with a tape measure attached to it. The bottom of the pole should touch the dock and the top should

point straight up. Have him turn the pole so that numbers on the tape measure are aimed at the porch. Then, get down on the porch and sight along the level to the pole, and read what number is in alignment with the level. If the pole is too far away, you might need binoculars or a gun telescope to read the number. It is important than the pole is held plumb when you get the number. Then have your helper let the pole down, and you measure the distance from the base of the pole to the number that was in line with the level. That distance is the rise to the porch, plus a couple inches for the height of the level on the porch. The tape attached to the pole needs to stay put, but it doesn't need to start at zero. It would be convenient to start with an even number.

Sometimes it is hard to sight along a level. It is easier to sight through two thin strings. You can set up a pair of level strings at the top of the rise, bring them into alignment, and see what point on the pole they match. In the example of the porch and dock, you could tie a string across the room, a comfortable working distance from the floor. Before hooking the other end on a nail, have a helper with a level tell you when the string is right. Set the second string 4 feet away (if you are using a 4-foot level) and make sure it is level with the first string, which won't be hard to do if the string is within the length of the level. When you get behind these two strings, which should be pulled tight, and sight with one eye past those aligned strings to the pole, you can get a pretty good reading of the rise. If you strap the measuring tape to the pole so it starts at zero at the bottom and strap the case to the high end of the pole, you can sight the rise without needing to measure along the pole. If you know the height of the strings from the porch, subtract that distance to find the actual altitude difference between the porch, where the steps will start, and the dock, where they will stop. If your pole is only 20 feet high, but the porch is higher than that, you will need to do the transit work in stages, and add the rises for the total. You could use a *rifle* in cheap transit work. Set the barrel level on a table at the sight of the top step, and fire a small bullet at the post (assuming your helper has propped the pole to stand by itself and is out of the way). Fetch the pole and measure the distance from the bottom of the pole (which was sitting on the sight of the bottom step) to the bullet wound. Subtract the height above the top sight that the gun was set at from that measurement to find the rise between those two places.

Whichever method you use, once you find the rise of the project and you have decided on a step height—say 10 inches—you will know how many steps it will take to climb that rise. Suppose the rise is 30 feet, or 360 inches. 360 ÷ 10 = 36. You would need 36 steps to get from the porch to the dock. If you want to use the porch and dock as steps, you would only have a rise of 340 inches and need 34 steps.

Start at the bottom with a tread depth that allows you to follow the slope of the land. If that slope is obviously less than 45 degrees from level, the tread depth will be greater than the altitude. If you wanted to keep them the same for some reason, you could do a series of the steep steps, followed by a level section, followed by the steep steps so that they stayed with the overall slope. You could use a set tread depth, and the steps can wind up the hill to remain uniform.

This is a lot of extra work because the extra "run" means more of the 10-inch steps. This work could be necessary if you find your steps too steep to step on. If you are willing to do the extra work, you can wind your way up a steep hill with fairly low steps having wide treads. The least work would be the straight, steep steps. If the hill is steeper than 45 degrees, however, you will probably need to wind the path.

DRY-LAID STEPS

Starting at the bottom, excavate a level area to accommodate the first step. Make the excavation go further back than the tread itself to get some overlap on the next step. Suppose you have decided on a 10-inch step with a 12-inch tread. The more you overlap the steps, the more stable they will be. At least a third of the stone should be overlapped, so a 12-inch tread would need at least an 18-inch excavation.

For a dry-laid step, you only need to dig down into the ground enough to give the stone a level seat. It is nice to get the lumpy stones and roots out so the stone will sit well. I like to leave the flat surface with a little layer of soft dirt, which will tend to fill in the little discrepancies and give the stone a good seat. Sand can also be used, but would be more likely to wash away on a steep slope.

Guides for Steps

If you have decided on a straight line of steps, it is helpful to have guide strings set along the ground as wide as the steps and parallel. While you are excavating, you will need to push the line out of the way to avoid cutting it with a shovel.

If you need to build a lot of steps and have a hard time with measurements and levels, but still desire a uniform look, you could use boards to make a form the size of your desired step. Suppose your steps were to be 10 inches tall, 36 inches wide, and 18 inches deep (counting the part to be overlapped). You could make a rectangle of that size with 10-inch boards, lay it over the spot for the step, and make a line in the dirt around it for the excavation. Then take it up and dig the hole until the form will sit level in the excavation. If you don't actually lean the rocks against wood, you could use the rectangle as a guide to set stones, and pull it off the step when that space is filled.

To get a better sitting dry-laid step, dig the excavation so all of the stone step will be below grade, but the lower edge will come to the surface. If you want an even more stable set of stairs, dig down so the form sits level and all of it is 4 inches below the ground level. Then pull out the box and pour in 4 inches of gravel as a quasi-footing for the step. The form is helpful in making a step because you can quickly tell if a stone is above or below the desired level of the top of the rectangle. You can minimize puddles by giving the entire step a slight outward slope to shed water.

If you use a wood form, be sure to set the rocks so their flat edges are on the perimeter of the form. In this way, when the wood is pulled away, there will be face stones on the exposed edge of the step. In other words, use the wooden rectangle as a convenient measuring stick and level, but definitely not as a brace for rocks that won't stay put. Remember to use your best stones where the stepping takes place. As usual, what you need is a big flat rock with square edges.

The part of the stone step that will be overlapped from above should not have a joint (space between stones) where the steps overlap. This would greatly diminish the way the steps are connected.

Options for Dry-Laid Steps

FIGURE 10-5 shows a possibility for uniform-height steps that vary in tread depth to accommodate a variable slope. Landings are used to give a break between small flights of stairs that have the same tread depth. In the drawing, starting at the bottom, the climber takes three uniform steps, has a pause in the pattern at the first level landing, then takes four uniform steps of a steeper angle and shorter tread depth, pauses on another landing where the pattern is broken, and takes five steep, but uniform, steps to another landing. This arrangement is more comfortable to the body than having a totally random flight where you have to pay close attention to each step.

It is best to get steps that work with one stone. If you can find big, flat stones, it is worth the extra hassle to use them for steps. FIGURE 10-6 shows a flight of stone steps that have been in use for a century. Over the years, the dirt has come out under the front face of some steps, and the space has been plugged with portland cement. The stones heave a little in the winter and the mortar breaks. Still, the steps are in use and very stable.

A set of stone steps of this nature with smaller rocks would probably have needed major repair by

now, but big, heavy stones just don't get knocked around very easily. The ones in FIG. 10- weigh over 300 pounds each, but it is worth the effort, and with help, really shouldn't be any more strain than working smaller stones. I think it is generally faster work to use monster rocks, because there is no fitting to be done within individual steps. These stones are strictly rubble; they were lying around the ground and cost nothing.

If your rocks are too small, building a dry stone stairs will be a problem. The stones would be so light, they would move from the weight and motion of humans. The worst kind of stones to work with, as far as your back is concerned, are the ones you can barely move. The ones you obviously can't move bodily will need levers, pulleys, rollers, chains, trucks, and teams of people, and they aren't as much strain.

Fig. 10-6. This old stairway has survived because of the size of the individual rocks.

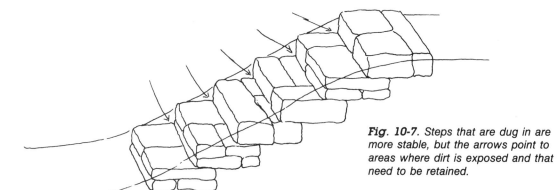

Fig. 10-7. Steps that are dug in are more stable, but the arrows point to areas where dirt is exposed and that need to be retained.

If you can't find stones that are heavier than a big person, flat on top, and not thicker than 1 foot, you have three options: you can put up with slightly shaky, dry-laid steps; you can bury your steps with the top surfaces of the steps at ground level, so the sides of the step are supported by earth and the smaller stones have nowhere to fall to; or you can lay the stone in cement mortar, attempting to make a big stone from smaller stones.

The second option is reasonable, but creates some new problems. The problem comes from making a level hole on a sloped surface. If the downhill end of the stone is to be at ground level, the uphill end of the stone will be below grade and more, leaving a ledge of dirt exposed that will need to be retained. For this reason, buried steps need to be wider than surface steps to accommodate the little retaining walls that must follow the steps on both sides. The arrows in FIG. 10-7 show the exposed sides of the excavation, which would need a stone retainer. This could be done with a very thin, low wall. If you are willing to grunt through the extra work, dry-laid stone steps of this sort will be nicely stable.

If you decide to take the third option of laying the smaller stones in mortar to create conglomerate stone worthy of a step, you have two new problems. The first problem is the cement. It costs, you have to get it, it spoils easily, it is hard to mix, it is messy, etc. The second problem is also the mortar. When it hardens around the stones, there is no flex when the ground moves from frost and it busts apart. The answer to this problem is to build a footing below frost line for the steps so they will effectively be sitting on ground that doesn't move. In places where the dirt freezes down 3 feet, doing steps with a footing can be double the work.

The frequently encountered tendency is to dig an inadequate footing, then lay the stones in extra strong mortar, and watch the steps bust and crack over the years. There is a certain merit to this method, sloppy as it might sound. You can use the mortar to level things and fill spaces, as you might use clay, with the assumption that the stuff will get broken by ground motion. If the stones are fairly large, the cracked mortar joints could be of little significance, except perhaps visually. With smaller stones, the mortar plays a more significant role in holding the step together.

It is also more important to have a good footing to keep the stones from moving. Concrete steps, with metal reinforcing, if they are poured as individual slabs, can ride out a frost heave and don't need as much footing. If you overlap each step by half of its depth, there will be a double thickness of step at all points. Steps overlapped that much will shed water better too, more like a shingled roof. More overlap, of course, means more gross tonnage.

Like with everything else, there is a right way and a wrong way to make steps. The right way takes about three times longer and lasts nine times longer than the wrong way, yet the "wrong way" is still a viable alternative. It would be very safe of me to mention only the finest approach to stonemasonry, but there is a large mass of people who can take advantage of less than perfect construction techniques. An unmeasured birdhouse without square corners or sanded wood will still serve the birds just fine. A stone stairway of less than perfect specifications can get you up and down.

Stairs through a Retaining Wall

Sometimes a long retaining wall needs an access. If the wall is 3-feet-high, you can build a set of steps into the wall without using much extra stone. FIGURE 10-8 shows such a situation in progress. We decided to use an independent stone for the first step, which can be put in last. In the wall, there is purposely a big, flat stone that comes to the height of the second step. Where the board rests, the wall is about 20 inches tall. The big flat stone on the board is coming off a pick-up truck and going directly on the wall. It will be recessed from the face of the wall the distance of the step tread and become the third step. The new step coming on will rest partially on the deep stone below. For added stability, these stones are also tied into the wall horizontally, with rocks on top.

A stepping stone needs to be extra long. The part that is covered might be 18 inches, counting both ends. This 3-foot-thick wall has capacity for three steps of 12-inch tread depth. The ground at the bottom and the top sport two more steps. As seen in FIG. 10-9, there is a flight of five steps here, with three steps actually being within the retaining wall.

Notice the need for side walls to retain the earth where the steps are below grade. It is best if you can find step stones long enough for these side walls to sit on them. Again, good dry-laid steps imply big stones.

FIGURE 10-10 shows another dry retaining wall with a flight of steps through it in progress. The man is building up the sides of the wall where the steps interfere. The first step is independent and very large. I like the option of being able to dig down a bit to get a thicker than usual stone to act as a step. You can't do that in this wall. The second stone step is tied into the wall at both ends, although marginally on the right side. Still, for that stone to be moved by a climber would entail moving a lot more stone that rests on it. Unfortunately, that second step is not set very deeply into the wall. The step above will barely touch it, but it is heavy enough to be stable anyway. The top rock on the right has its long side going into the wall. This makes it easier to construct the sides of the steps. These steps are shown completed in FIG. 4-34, lower right.

STONE STEPS IN MORTAR

FIGURE 10-11 shows a two-step arrangement that is part of a stone wall. Each tread is 12 inches because the wall is 2 feet thick. If I had wanted to get three steps in the wall, I would have been limited to an 8-inch tread. The altitude from the floor to the doorway would have been too much for one step, and a step set on the floor would have taken up valuable hallway space. Notice that the large, flat stone at the bottom has been brought up to the step height with other stones. One of those stones ties into the wall to the right. This situation must be somewhat preplanned or anticipated, as does putting a capstone on a wall. A stone step is a capstone of sorts. You should pick out your step stones early and reserve them for the occasion.

Measuring Stones

The stones in FIG. 10-11 are so uniform it was easy to measure for them. With rougher shaped stones, it is harder to calculate the amount of space they will take up. From what I've seen, people measure rubble too small. To get a good measurement of the maximum thickness of a stone, turn it on edge, hold it plumb, and eyeball straight down on the rock with a measuring tape over it. If you see any protrusions, you must figure them in, unless you knock them off with a hammer.

If you have a stone with a very uneven bottom and you can't predict how much lift it needs to become a step, you can make a very deep mortar bed and push the stone down into it until its top has come to the right level. You will need to be ready with a trowel to catch the mortar that squishes out. If the mortar joint that remains is too big, you would need to take up the big stone and lay in some thin ones to bring it up higher. Sometimes the small flat stones can be shoved right into the mortar joint, and they will jack the step some without needing to lift it off and on.

It is wise to get your distance figured before you lift the step stone into place. Pulling a big flat stone off of a wall makes a big mess, and it is heavy work, even with help. The bottom step stone in FIG. 10-11, which goes into the wall under the second step, weighs 250 pounds.

Moving Big Stones

It helps to have lots of people on hand when you want to move a big stepping stone. The real problem is that the stone must be lowered onto the wall, if there is a bed of mortar. If you try to tip a big rock on edge in a bed of mortar, the mortar will instantly squish out

Fig. *10-8*. *A big step stone is moved on a plank directly from truck to wall.*

Fig. *10-9*. *These steps penetrate a dry retaining wall.*

Fig. *10-10*. *The bottom step here is not connected to the wall. The second step is tied in horizontally to the wall.*

Fig. 10-11. These steps go through a mortared stone wall.

the front and smear the face of the stone. The mortar toward the rear of the stone will mush into a wedge shape. Tipping a jumbo rock onto a wall can also make a mess of the freshly laid stones below. It is too much weight all on one point.

The dilemma is how do you lower a big, flat stone onto a wall? When there is a wall in front of you, it is nearly impossible to use proper body mechanics. This is why rubble masonry can be so devastating on the spine. Unless you can somehow get on top of the wall, there is no way to avoid holding the big weight out away from the spine.

There are various devices that can help you move big rocks. If you can rig a steady tripod that holds a winch or come-along hooked to a chain and you can move the rig along the wall, it will help a lot. This is not always possible, however, and when it is, it is difficult to keep the chains or ropes out of the way when you lower the big rock.

In the old days, a big cut stone was moved by making a wedge-shaped hole in the top of the stone (the hole gets bigger as it goes into the stone) and inserting a matching piece of metal with a hook. The metal was in two pieces so that it could be put in the hole. A chain would be hooked on and attached to

a crane. The stone could be lifted and lowered into the wall without any obstruction underneath. The metal wedges would tighten in the hole when the weight was on. After the stone was in place, the metal pieces would be hammered into the hole, where they would have enough room to be slipped apart.

I mention this method as a curiosity more than a suggestion. Your best shot at moving big stones is to walk them up a ramp and get help to get them in the right position on the wall. If you think the thing through, you can position a big stone so that its last flip will drop it in the right spot.

It is good to have some extra hands to gently lower the rock on the mud bed. The ramp board should not sit on the wall, rather on a temporary pier or brace against the wall and at the correct height. If the board is actually on the wall, you could have several problems (FIG. 10-12). One, the weight on the edge of the wall could tip up the stone the board rests on. Two, the board could come flipping up when the weight of the stone gets on the part that is unsupported. Third, the board could get stuck under the stone when the rock is finally on the wall. For dry walls, or when using smaller rocks, putting the ramp right on the wall is probably all right.

Fig. 10-12. When ramping a big stone onto a wall, it is best if the board doesn't touch the wall.

When several people are moving a big rock, it is important to communicate and have good timing. It is easy to get a finger crushed if one person moves before another is ready. Don't get in a hurry with 300-pound stones and you won't have one fall on your toes. Make sure the path is clear. You don't want to be tripping on a roller skate or a sleeping dog. If you need to get a big stone through a doorway, make sure there is enough room before you get the rock in the air. If it is going to be a tight squeeze, make sure everyone's fingers are out of the way. It is horrible to discover, after the big rock is in motion, that your hands won't clear the opening.

A wheelbarrow is invaluable in rock moving, although it has its limits. The top stone in FIG. 10-4 was lowered on a heavy-duty wheelbarrow and immediately popped the tire. For a good wheelbarrow, 300 pounds is about the limit. Don't even bother with a light-duty wheelbarrow with a solid tire. A full wheelbarrow is very unwieldy and it is good to have one or two people steady the load. If the wheelbarrow must go through any openings, make certain there is enough width to accommodate the knuckles of the driver. All too often, a little skin is left on the door jamb. The momentum of a big stone is too much to stop or turn on a dime. Don't try to bank, or lean in a turn, with a full wheelbarrow; it will tip over. The wheel needs to be straight up and down to stay balanced with a big load.

Using Small Stones

You can build mortared stone steps with small stones if that is all you have. The rocks need to be very clean, and you should add a little extra portland cement to the mix to ensure all the bond strength you can get. A dusty stone will not bond with the cement at all.

Good curing temperature and moisture are necessary for the adhesiveness of the mortar to be of much value. It is good to have a somewhat wet mortar when you want maximum adhesiveness. The wet mix will squish into the smallest spaces between stones. The problem is, it will also squeeze out of the wall and down the faces of stones below. If the work below is already hard, you can wash the drippings down the wall right away with a fine mist from a hose. Using white portland cement instead of the gray will lessen the staining from mortar drips. If the work that is dripped on is not set, you will need to wait up to one day before scrubbing the mess off the wall with a stiff bristled brush and water.

Stone steps don't need to intersect a stone wall (FIG. 10-11). They could merely abut the side of the wall. Stone steps are special, however, and deserve the best you can do. No other rocks in your life will serve you as much or take as much abuse as stone steps. You can get away with a lot of rule breaking in rubble work, but don't skimp on steps. Bad steps are worse than a bad wall.

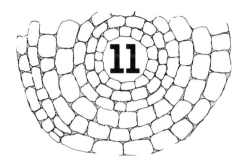

FLAGSTONE

FLAGSTONE is not a type of stone, like granite, limestone, or sandstone. Nor is it slate, which is a metamorphic shale. Flagstone is a shape of stone. It is any stone that is thin and flat, like a flag. That thin, flat stone can be any type, although granite flagstone is unlikely. Flags are often cut at quarries into uniform sizes with square corners. It also can be bought as rubble with irregular shapes. In Pennsylvania, most of the rock that has been quarried for flags is bluestone, a kind of sandstone.

Usually, stone suitable for flag quarrying is sedimentary, with distinct layers that separate into slabs 1 to 3 inches thick. The stone has to have a tight texture and be hard enough to withstand traffic if it is to be used as finish floor. Shale comes apart in layers, but it is too soft to use as flagstone. Slate can be used successfully as flagstone, and often is, although it might be harder to get and more expensive than thin pieces of locally occurring stone.

FINDING STONE

In some areas, you can find thin, flat stones in creeks that are okay for a patio or walkway outside. It is unusual to find rubble stone in the right shape for an interior floor, however. Big, flat pieces occurring naturally tend to get broken over the years into smaller pieces.

It is quite possible to find a ledge of sedimentary stone that is exposed where you could quarry your own flagstone. (Get permission first from the owner of the land.) It might be easier than you think to break apart the horizontal layers. Start at the top of the outcropping and try to pry off the top layer using wedges in the crack between layers. It is best to hammer several wedges at the same time to get a big piece

to break loose. In a surprisingly short time, you can get enough stone to do a big floor. When your stonework is 1 inch thick, instead of 24, the gathering is fast.

One ton of 1-inch-thick stone could cover 160 square feet of floor. If the flagstone is 2 inches thick, one ton will cover an area 10 feet long × 8 feet wide. One ton of 4-inch-thick stone will only cover an area 10 feet × 4 feet. These figures are approximate at best. If your mortar joints are huge, the stone will go further than if you fit the stone very tightly. If your pieces are extremely irregular and parts need to be broken off and wasted, your ton won't go as far.

THIN STONE

If you find stone less than 1 inch thick and you would like to use it horizontally, you should refer to floor-tile instructions and lay the stone as a tile. Slabs of stone thinner than 1 inch must have a very uniform surface; they can't have much of a bump. This type of stone (most likely slate) will not be well suited to a rustic look or an inexperienced builder's trowel skills. If any pockets of air are trapped under thin stone, it will crack when walked on. Thicker flags are more forgiving of this very common problem because they have adequate tensile strength to bridge the small gap.

Thin stone implies a level of refinement in mortar. When mortar joints are very thin, they behave differently. At some point, a shift to fine, sifted sand must be made. If the joint was no thicker than the largest aggregate, it would be separated at that piece. The largest aggregate should be about one-third the thickness of the joint. Fine sand has more spaces than large and needs more cement paste for similar strength than regular sand. A very thin joint that needs much adhesion must be used as a *grout,* or liquid mortar. The extra water allows the mortar to flow into the tiny spaces, but it also weakens the mix, and consequently even more cement is needed. A mortar with fine sand might be one part masonry cement and two parts fine sand. A grout with fine sand might be a 1:1 mix.

At another point along this journey from gravity to glue, additives are put in the mortar or grout to improve the stickiness so the stones or tiles won't pop off. As the masonry gets thinner and thinner, its mass can't be considered as much of a stabilizing factor. Mortar mixes for terra-cotta floor tiles have sand that is crushed to unnatural fineness. Beyond this in fineness is *cement butter,* which is portland cement and water with no sand. This butter will really hang on to a stone. It is sometimes used to give a flagstone a surface that will grab the mortar bed better.

As the role of stone runs in the direction of veneering, it leaves the realm of the stonemason and enters the realm of the tile layer. The two have very different methods and skills, tools and mixes. In my opinion, a stone tile should be dealt with as a tile.

There is an area of overlap between the two crafts: rubble flagstone more than 1 inch thick. The irregular brown and green sandstone that I use is four times cheaper than similar stone with square cuts. I don't think there is any absolute method for laying flagstone floors. I have read about, seen, and tried different approaches. I have seen many flagstone patios that were done ''by the book'' but still busted and cracked. There is an inherent weakness in this kind of masonry that you don't have in a wall.

Flagstone is pretty new as a type of masonry. Before the advent of concrete slabs, stone floors and walkways were made of *pavers,* which were quarried stone of very hard surface cut into blocks at least 6 inches thick. The stones aren't much larger than a brick.

With stones that small and relatively many mortar joints, the floor can handle motion without busting, spalling, or cracking. The many mortar joints act as shock absorbers and control joints to handle expansion and contraction. Large flags are much more likely to break from a push below or from fast expansion from being warmed by the sun. If you lay big flags in mortar rich in portland cement, which is tempting to do to increase adhesiveness, you get a mortar that is as hard as the stone and doesn't give at all in stress.

USING FLAGSTONE IN FLOORS

The old approach to stone floors was the same as the old approach to stone walls. It is still far superior to using thin flagstone. Using flagstone horizontally and vertically is a compromise, not an advancement, in the craft. Nobody can afford a real stone floor, so we invented a way to use thin sheets of stone. No matter how you do it, though, it won't work as well as a thick stone floor. Still, a well-done flagstone floor should last as long as the house it graces, so there is no particular reason not do it.

The only kind of flagstone job I would avoid is one done on top of a wooden floor. I think it spells trouble no matter how you do it. A wooden floor flexes, expands, and contracts at a different rate than masonry. When the floor gives and the stone doesn't, the stone cracks. Another bad point is that the wood will rot faster with the stone on top because it can't breathe. Humidity will get trapped under the floor, especially if the floor is over a basement. The extra weight isn't great for the wooden floor either. The idea of having masonry depend on wood is backward and worth avoiding. Perhaps the best place to use flagstone is on the bare dirt, without mortar.

FLAGSTONE WITHOUT MORTAR

The most effect for the effort that you can get with stone is a flagstone patio in the dirt. It costs nothing more than the cost of the stone. Suppose you want a patio outside of a ground-level opening where there is too much traffic for grass to survive. If the area is fairly flat to begin with, it won't take much excavation work. On sloped land, a level patio would entail much excavation and a retaining wall. If there is lawn where you want the stone patio, you will need to till it up to break the sod. It is good to have a soft, pliable, fairly level, and rock-free dirt surface to lay the flag stones on. Don't till any deeper than you need to get a level patch that you could press a stone into, and get a nice, even backing.

Using a Water Level To Get a Level Dirt Bed

A homemade water level is so easy to make and so much fun to use, every serious weekend putterer should know about it. It is a handy device for bringing unconnected things to the same height. A water level is also good for establishing slopes. You can't develop an even slope if you don't know where level is. The heart of the water level is a length of clear plastic tubing with a small opening. Size isn't crucial, but a small tube is quicker to fill and has less effect on the water level in the reservoir. The small-diameter tubing is cheap and it is good to have about 50 feet. The other part of the water level is the reservoir. A 5-gallon bucket is fine for this. A smaller container will also work, but it might not be as accurate because the amount of fluid in the tube becomes too high a percentage of the total and if any is spilled out of the tubing, it would have a greater effect on the reading than would the same amount spilled from a larger container.

Find a place on your dirt patch that you think is about average level, that is, not the highest nor lowest place on the future patio site. With the help of the water level, you can push dirt around until the whole area is on the same level as the control spot where you set the reservoir. Set the bucket of water on a cement block at the control spot. Make sure that both the bucket and the block are stable. Fill the bucket to near the top and add some food coloring so the water will show up better in the tube.

Next, fill your 50 feet of clear plastic tubing by setting one end in the water with a weight to keep it in there. A brick will do if it doesn't pinch the tubing. Take the other end away from the reservoir and suck through the tubing with the end kept down lower than the bucket. The water should quickly come squirting out. You need to *bleed* the line, or make sure the tubing is completely filled with water and not air bubbles. With this done, pinch the end of the tubing closed, or hook the tubing to a tree or wall so the water won't spill out. You will only need to get the end of the tube as high as the top of the bucket.

With the tube full and a weight already in the bucket, mark the water level on the inside of the bucket with a crayon to keep a check on things in case a thirsty dog comes by and drinks 2 inches of your reservoir. Then you will be able to fill it to the correct spot again. Measure straight down from the water level in the bucket to the dirt. If that measurement is something difficult to remember, like 23 9/16 inches, pour a bit more water in the bucket to get it to 2 feet and make a new mark.

If the water in the bucket is 2 feet above the desired patio height, then the level of the water in the tubing, as far as 50 feet away, is also 2 feet above the desired level. The reservoir becomes a reference point. The water in the tubing, which is held above the level that would allow water to spill out, is at the same level as the reference.

Measure from the water level in the tube to the ground directly below. If it reads 22 inches, then you know the ground is higher there than where the bucket is. You will need to dig 2 inches down in this area. Suppose you walk the tube and your measuring tape to another spot and it reads 27 1/2 inches from the water level to the dirt. This part of the sub-patio is lower than the control spot. You will need to add 3 1/2 inches of dirt to this spot. If the water reservoir is centrally located in the job site, you can get by with the shortest tube. It is best to fill the low spots with the dirt from the high spots; The water level will quickly tell you which is which.

You can minimize your dirt moving by finding the average altitude. If your first reference spot shows too many places higher when checked with the tube, it is too low a spot to go by.

If you have a certain level you know you want to bring the patio to—for instance, the height of an existing patio—then set the water level reservoir on that surface, and measure down from the water top to the patio level. Keep that number in mind. When you check the adjacent areas you will know how much dirt needs to be added or subtracted to get the whole job site to the desired height.

Irregular flagstones laid in the dirt shouldn't be very large if you don't want them to crack. Pieces of about 1 square foot will ride out the motion of the ground much better than larger pieces. You also would want to use thicker stone on the dirt than what you would need on a concrete slab in mortar. Two inches should be thick enough.

If you want your stone patio to be square, round, or in any way uniform, you will need some guidelines. Remember, though, that lines are guides, not dictators. If you have a big curved stone that you don't want to cut, let it protrude beyond the string. When you're working with random stone, you might as well exercise your freedom to make crooked lines. You won't be able to make very straight lines anyway, unless you cut the stone. If you are very concerned with straight lines, fieldstone will frustrate the heck out of you. Every rock is different. You need brick and tile for uniformity. The irregularities of rubble should be used to advantage. Fieldstone is actually helpful for building unique structures. Stones don't dictate anything but weight and endurance. They begrudgingly fit into rectangular schemes.

Layout

After you put up guidelines, which are mainly used to see how crooked you are going, pick out your largest stones with a straight edge and set them on the perimeter of your site. Find stones that have long sides that match edges of the perimeter. Start filling in the smaller spaces with naturally fitting pieces. Treat it like a big jigsaw puzzle. I don't see that it matters much where you begin a flagstone job, as long as you don't need to walk on the fresh stone as you work. You could start at the center and radiate toward the perimeter, or you could start at the corners and work toward the middle. It makes sense to work away from the finished product, as you would if painting a floor. I usually start at the corners if it is a rectangle, and then fill in the rest of the perimeter. Starting at the straight exterior with the available straight pieces will minimize cutting on a floor job. If you worked from the center outward, it would be unlikely if your stones just happened to end on a predetermined borderline.

I like to do a dry run with flagstones. This is too much extra work with a stone wall, but on a floor, a ton goes so far that it is worth the hassle of lifting the rocks an extra time to do a trial fit. For a stone patio in dirt, I wouldn't bother cutting stones to fit tightly or keep uniform mortar joints for the dry run. Fit them as closely as possible, always keeping a space around each stone.

After you have a dry-fitting arrangement that works, move your stones off the patio space and carefully arrange them in the same position adjacent to the work area. To reduce confusion, keep the stone in the same orientation. This relaying can be very fast and sloppy, as long as the relative positions remain the same and can be copied.

With the stones out of the way, you can begin laying the finish surface. Assuming your dirt is a little fluffed up on the surface, with the rocks and bumps removed, push your first stone into the dirt. It should feel solid and not wiggle. If the rock wiggles, pull it up and look at the depression in the dirt. You can easily see the hollow spots, and you can fill them with more dirt. For a better adhesion, make the dirt wet before you lay each stone. You can push a rock into the mud, and it is hard to pull it back up. If places in the dirt have become too compressed from walking on, you might need to loosen up the dirt some with a hand trowel before you can get your stone to nestle in the ground. Using water will help a lot, if you can keep it out of your way while you work, as you would with mortar. When you wiggle a flagstone into the soft clay, make sure all of its edges are below grade.

Keep the water level set up, and check your work as you go to make sure it is level or sloped the way you want. More important than level at some point in the progress is that the edges of the stones don't stick up. The stones should be flush with each other, and a long straight edge, like an 8-foot 2 × 4, is more helpful than a level to keep stones on the same plane. If you use a straightedge, make sure it has a straight edge by sighting along it. If the edges of some stones protrude, people will catch their feet on them. A hump in the middle of a stone is more tolerable, because it presents no abrupt edge to the toe.

Sound is helpful when laying flagstone. By tapping on a freshly laid stone with a trowel handle, you can hear if there is a hollow under the stone. If there is, pull up the rock and add more dirt to give even support. Stones under a flagstone will cause it to break, so if you feel any stones while you lay the patio, toss them out and fill the space with more dirt.

Sand is often used to lay dry flagstone. I think dirt is better for several reasons. Foremost, the dirt will support a joint of vegetation (grass), which not only holds the dirt, but also the stones, in place. Sand, on the other hand, tends to wash out and won't support the growth.

After getting all your flagstones pressed down into the moist, soft earth, fill all the spaces between stones with smaller stones, then fill the remaining spaces with dirt. A mason's trowel is handy for this job. In fact, a mason's trowel is a darn handy garden tool, so if you garden, you might find the cost more bearable when you buy one.

After your stone is down and the spaces filled tightly, clean the dried dirt off the patio with a broom and dust pan. Then plant a fine grass seed in the joints. When the grass takes hold, it will pinch the stone to the ground. It is surprisingly hard to remove stones surrounded by grass. The grass will also buffer the stones so that if there is movement, stones won't crack; they will bang against a cushion of dirt and roots.

If you pressed the stones into the dirt as you went, there shouldn't be a whole lot of settling afterward. It is surprising how little compressive force a body exerts on the soil when the weight is distributed by a stone of 1 square foot or more. Suppose you weigh 144 pounds, and you have each foot on a stone of 1 square foot. You would be exerting a force of 1/2 psi on the dirt. The compression will be less than if you walked on the dirt. Unless you plan to drive a car on the patio, there's no reason to compress the dirt a lot beforehand.

When the grass appears in the cracks, which could constitute one-third of the whole surface, you will need to mow, which won't be a problem if the stones are flush with the dirt.

There is not much history of patios done this way, so I cannot make any juried claims to the longevity of the arrangement. If something would go wrong with a patio of this sort, I do know that it would be easy to fix. Repairing people's portland-cement fantasies is a real mess. Patios done the "right" way do break, and they are hard to fix. You don't have to look too far to see the evidence of the weakness of laying large slabs of hard stuff on the ground. Look at the sidewalks in towns and cities. Look at the driveways in the suburbs. There is no end of the busting concrete that was done by the book.

We tend to think of concrete as permanent, but it isn't. Many masonry buildings and floors last no longer than a good frame house. The so-called ageless masonry is the old, very conservative, time-proven stonemasonry. Flagstone isn't in that realm no matter how you do it. The downfall of the patio on the dirt would probably be furniture with small feet. The weight in such a condensed area would possibly break stones.

LAYING FLAGSTONE IN MORTAR

When you lay stone in hard mortar, it must have a footing that carries the load down to firm, unmoving subsoil. You could lay stones right on the lawn and mortar them to look official, but the stones would bust apart as soon as they moved—and they would move. Trees will send roots that can move the stones. Water will make ice and move stones. A dry stone patio on a surface that moves is a flexible mesh. A mortared patio on ground that moves is like a sheet of glass on a bed. A mortared patio on a firm foundation below frost line is like a sheet of glass on a concrete floor; you can walk on the glass without breaking it.

Even on a good footing, irregular stone floors will crack if they go through a lot of expansion and contraction. On a large patio, an expansion joint is a good idea. In lieu of that, a soft pointing mortar will allow some expansion without stones breaking. It is always best to have the mortar softer than the masonry units. If the mortar joint breaks, you can dig it out and replace it in a few minutes for a few cents. Good masonry is designed to be maintained, but needs little of it.

A Mortared Floor Inside a House

A laid-in-mortar flagstone floor inside a house must be done on a slab of concrete or something else that doesn't bend. Don't attempt to lay stone on top of foam insulation, unless the foam is under the slab. The little bit that it would give would cause chaos at the surface. I've never heard of this being done, but a flagstone floor could be laid on compacted subsoil, if it was below frost line and good drainage was provided at the perimeter of the foundation. A rammed earth floor would also work with stone on top.

I have seen beautiful interior brick floors laid in sand. The sand bed must be on top of something solid, like concrete. The bricks are set in the sand without spaces—tightly against each other. When they are all in, fine sand is swept over the surface and into the spaces. A floor like this works, even in an elegant house. With flagstone, the same procedure probably wouldn't work well, unless the stone was as thick as a brick and could be fit tightly.

I encourage experimentation in masonry, but don't want to get sued if you try something that doesn't work. An area I think deserves investigation is that of alternatives to concrete slabs for below-grade interior floors. Concrete makes a miserable interior floor and isn't as cheap as you might think if you count the labor.

Pouring Concrete

If you decide to lay flagstone on concrete inside, which is what most people do, take the extra time to get the slab level (or evenly sloped). The time spent on the concrete layout will save you time when you're mixing mortar by hand to lay the flagstone. The more crooked the slab is, the more mortar you'll need. If you get the slab how you want it and if the flagstone is all the same thickness, you can lay the stone floor without any guidelines. You just follow the floor and keep the edges of the stone on the same level. Use the thinnest bed of mortar that will comfortably seat the stones.

If you do pour a slab with the intention of laying stone (or brick) on top, don't power-trowel or hand-finish the concrete. Not only does this take the most skill, extra money, and time, it weakens the top layer of concrete by bringing the water and cement to the surface. This leaves the concrete just below the surface depleted of cement paste. A smooth finish on the slab also will prevent the mortar bed from bonding with the slab. A rough-textured, but uniform slab will give the mortar more tooth and save you money.

The slab should be of uniform thickness, mostly to use the least materials. Four inches is typical. If you use a lot of metal mesh and smaller gravel, you could make a thinner slab.

The easiest way for a beginner to pour a good slab is to set guide boards in the gravel under the slab. The guide boards should be straight as an arrow, and you should set the top side to the desired height of the slab top with a transit or water level. These guide boards, usually long 2 × 6s, run the length of the slab on both ends, and every 10 feet or so of width. So if your slab was 20 feet wide, you'd need three guide boards: one on each end, and one down the middle. Take your time to get the guides to come to the right altitude. When you get down and sight across the boards, they should all fall in line. When the concrete comes, you will use these boards to ride a screed board along. The screed board is another very straight 2 × 4 or 2 × 6. It needs to be longer than the width between the guide boards. If your guides are 10 feet apart, the screed should be 12 feet.

Suppose you were going to pour a slab 16 feet × 20 feet. It is customary to put 4 inches of gravel under the slab. To calculate the amount you will need, you need to know the size of the vacancy. Four inches is 0.333 feet. $16 \times 20 \times 0.333 = 107$ cubic feet. Concrete yards deal in cubic yards. There are 27 cubic feet in a cubic yard. $107 \div 27 = 3.9$ cubic yards. Call it 4 yards.

After you get the 4 yards of gravel spread evenly, you need to get a few helpers with a strong rake, shovel, and rubber boots. Because the slab is also 4 inches thick, you need 4 yards of concrete, same as the gravel. The concrete will weigh 8 tons.

If you have good access to the job site and the truck can get its chute right where you need it, you won't have to lift those 8 tons. Start at one end with a helper and have the driver slowly unload the glop between your guide boards. When it looks like the area is full, start to drag the screed board along the tops of the guide, with one person on each end of the screed. Jiggle the screed board vigorously back and forth as you push it forward. The jiggling helps to spread out the concrete. Make sure the screed board is held tightly against the guide boards.

The guide boards need to be staked into the ground to withstand the abuse of the concrete. A hole through the top of the 2 × 6 will allow a section of 3/8-inch rebar to be hammered into the dirt. This will hold the board in place without interfering with the concrete on the sides or the screed on top.

As you push and jiggle the concrete load along, watch for high and low spots. You will frequently need to lift the screed and go back over the same area to get the mix to fill all the vacancies. If too much concrete builds up, you will need to shovel some of it out in front quickly. It takes awhile to get this right. Don't get too hurried. You probably have an hour. After that, the concrete will be getting too stiff to do anything with.

When the concrete has stiffened enough to walk on, pull up the guide boards. This will leave a 1 1/2-inch gap in your slab, which you can fill by hand later, or fill right away with a wheelbarrow-load of mix set aside for this job. This is a good opportunity to add control joints to the slab. When you fill a gap, leave a space with a strip of cardboard or a couple layers of tarpaper. These control joints are a way of breaking the slab before it breaks itself. The slab will have somewhere to expand to.

This is not a book about concrete work. For more detail on that part of a flagstone job, you should read something on concrete and have a helper who is familiar with it. The technique I mentioned here is for the novice. It will give good results, provided you intend to lay stone on top. A slab that is merely screeded and not troweled at all will make a rough surface for an interior floor.

You might wisely wonder if you could lay the flagstone directly in the wet concrete, and save the separate step of mixing mortar for the stones. It has been done, and I think the problem is that you don't get enough time before the concrete sets to fit the stones very well. There is a tendency to race through the fitting job, and then you have to live with an ill-fitting floor. An inside floor especially should have care taken to get stones to the same level. A rough floor inside can be a hassle to live with. If you were mixing your own concrete and doing a small area at a time, you could lay flagstone in concrete, although the concrete would be an unwieldy mortar if you were trying to wiggle a high stone down some. It is best to use a smoother mortar when laying flagstone.

When laying flagstone in mortar, you must plan your procedure because you cannot step on the freshly laid work. You would do as painting a floor: work away from the work toward a doorway. Inside a house, you will probably want tight-fitting stones and semiuniform joints. So you will probably need to cut some of the irregular flags. Before you cut anything, though, it makes sense to put in the big pieces as they are, probably with a straight edge on the perimeter. If you have any pieces with a square corner, put them in the corners. Work in the new pieces by finding a side that fits well with one of the edges already laid. The more natural fits you can find, the less you will need to cut. If a stone has a big protrusion which makes it difficult to fit next to anything, chop it off with a sharp hammer blow. Try not to create difficult areas to fill. The shapes you isolate with stones should be fairly likely shapes, not zig-zags and amoebas.

I recommend you do the initial stone fitting and cutting without mortar. This way there is no urgency imposed by the setting mix. You get time to fool around with the shapes and the fits without the worry of the mortar. After you complete the dry run, you can take the stones outside and set them in a similar rectangle on the lawn, one by one, in the same position. This isn't as much effort as it might sound like.

As you dry-lay the flags, pay attention to the overall look. If you have two different colors of stone, intersperse them. Don't use all your big stones in one area. Don't make one section mostly triangular and another mostly square. Always leave a space for the mortar.

Eventually, you will have to cut the stone to fit. You can use a Skil saw with a masonry blade to score the stone quickly. Then you can break it with a hammer. You also can use the saw to cut all the way through the stone, to prevent wrecking a precious piece with a faulty hammer smack on a shallow score line.

Wear goggles and a face mask when you are cutting stone. The dust that comes up is incredibly fine and bothersome. It will get in the saw bearings, so don't use your favorite tool for this job. You can't really turn a sharp curve in the stone with a saw. Don't expect to cut the shapes of the 50 states in flagstone. You could do Kansas and Colorado, but Maryland would be tough.

When you have isolated a space with stones, you can take a rock from the selection pile and hold it over the space to check the possibilities. Try to align a natural line on the stone with a line in the vacancy. Where the stone overlaps the space, mark the end of the space on the rock with a pencil. Cut parallel to that line 1 inch inward to allow a mortar joint. When you cut the stone, first make a quick shallow pass. If you go in deep right away, you can't steer the saw. The wilder your line is, the less chance of the stone breaking along it. Break the stone in a bed of sand to provide support and shock absorption for the good part of the stone. The deeper you score the mark, the better chance of the stone breaking there.

You can also score flagstone with a hammer and chisel. This is slow and the breaking is even less certain, unless you score the line deeply on both sides of the stone. The idea with scoring a stone is to make an obvious weak line. If the stone already has a major fault line or crack, there's not much use competing with it. Cut stone in the direction of the grain. Cross-grain breaks won't work well, unless you saw clean through the rock. If your sedimentary stone appears to be in layers that would come apart, they probably will when you try to break the stone. The ideal for flagstone is to find a sedimentary stone that is in layers 1 to 3 inches thick, and that won't delaminate any further.

If you can't tell where to make your cut line because the new stone is bigger than the space and won't allow a view of the space, you can make a quick template with aluminum foil. Press the foil into the space, peeling the edges back enough to make a mortar joint. Then pull out the foil stone and lay it on the stone that needs to be cut. Trace the pattern on the stone, aligning as much of the shape as possible with the actual rock edge to minimize cutting. On dark stone, a nail works better than a pencil for marking the cuts. This is an easy way to fit an awkward shape. You can use the same piece of foil over and over.

When you have all your stones in place on the dry run, move them aside, keeping track of positions. Now you're ready to lay the stones in mortar. Start in a corner and lay down a bed of mortar thick enough to take up all the indiscrepancies in the rough concrete. It should be about 1 inch thick, and spread on a clean, moist slab. The mix should be pretty rich for adhesion. One part masonry cement to two parts sand will do. The mortar should be moist, as should the stone before you lay it. A dry stone or slab will absorb moisture from the mortar before it has had a chance to set.

Lay out enough mortar to go slightly beyond the area of your first few stones. Set the stones on and wiggle them into the mortar for a good bond. Don't let the mortar squish out on top of the stones. It is all right if it starts coming up the cracks, but don't try to point until later. Spread out some more mortar in a roughened up bed, and lay a few more square feet of stone, making sure the edges are not protruding and the tops are flush with each other.

FIGURE 11-1 shows the order that the stones might be worked in on the dry run. As you work stones along the perimeter, you will eventually meet the corner piece you selected first. It is unlikely to find a stone that will fit that isolated area without cutting.

The illustration shows that the No. 10 stone laid has come to meet No. 5. Until then all the lengths and widths of stones had been arbitrary. The No. 12 stone must be a particular dimension to fit on the perime-

ter. You could get lucky and find a stone that fit that space. It is worth looking through your pile, and if nothing else, it will help you find a piece that is just a little too large that you can cut for the spot. In the drawing, the No. 11 stone was chosen because it fit the curves presented by Nos. 10 and 5. Number 11 was put in before No. 12 because once it is obvious you will need to cut a stone like No. 12, you might as well cut it to a specific size that will nicely accommodate a secondary fit like No. 11.

At this point, rather than build more in that area and block my access, I went back to the far perimeter for stone Nos. 13 and 14. The No. 15 stone, like the No. 12 stone, had to be cut to fit because it was hemmed in between the No. 2 and No. 14 stones. Again, find the stone that nearly fits and cut some off.

Avoid filling the space with smaller stones. The perimeter of a floor is like the cap of a wall; it is the container and those stones should be the least likely to move. The weight of a 2-inch-thick flagstone can be a structural factor. A piece 2 feet × 3 feet × 2 inches would weigh 150 pounds or more. A stone like that has a lot of tendency to stay put regardless of the adhesion of the mortar. A small stone, weighing only a few pounds, might get kicked off the edge, especially if there is an exposed ledge and you are treating the flagstone as a step. Only the perimeter stones would ever experience weight on their edges, which could flip them up (FIG. 11-2).

The idea of crossing joints on a floor is mostly a visual thing. Continuous seams with irregular stones look bad because they aren't quite straight, but they aren't random either. FIGURE 11-3 shows the order I might go in after the perimeter pieces are in. The doorway perimeter has been left out to maintain an access. I kept looking for pieces that match sides with ones already in the floor until an isolated area was created, then I cut a stone for that place. I placed Nos. 17, 18, and 19 and cut No. 20 to fit between Nos. 19 and 3. Then it was back to the far end.

Stone No. 21 is a lucky fit of the type that comes from watching the shapes. I cut No. 22 to avoid a future problem. I chose No. 23 for the nice fit against Nos. 14 and 15. Remember, when No. 23 was put down, the higher numbers around it weren't there, so there were no restrictions on its dimensions except to mesh with Nos. 14 and 15.

Continue the process of putting naturally fitting

Fig. 11-1. The order you might lay flagstone on the dry run.

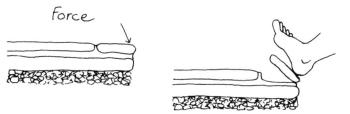

Fig. 11-2. Small stones on the edge of a patio can come loose.

Force

edges together until small, isolated areas are created and filled with custom-cut stones. Lay out the stone to avoid continuous lines.

As you near filling in the area on your dry run, look at your rock pile. Now is your chance to make sure that the best of what remains gets into the floor. If you do it right, you should end up with some extra pieces—the worst of the smaller stones, plus some pieces of cuttings. You can use some of the cuttings to fill unduly larger mortar joints, but they must have a space around them too. None of the stones should touch. If a fill stone is as wide as the joint, say 1 inch, it will look fairly bad. You should avoid it by more careful fitting of larger pieces.

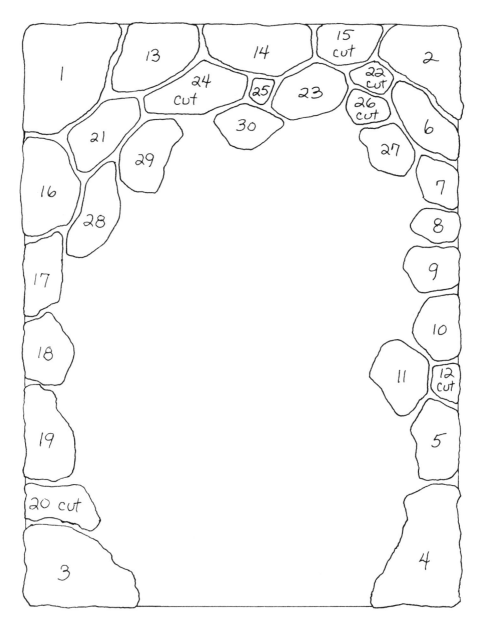

Fig. 11-3. Follow a logical procedure when laying irregular stone to fit, to avoid cutting as much as possible.

Following is a cheap trick to get some perfectly fitting flagstones. Lay a big stone in a bed of sand and smack it in the center with a sharp hammer blow. It will break into two or more randomly shaped pieces that fit each other perfectly. You can expand that arrangement to include a mortar joint between pieces. If it looks too unlikely, though, it will stand out poorly among the humble neighbors. A sense of continuity is essential for the maximum aesthetics in any kind of stonework.

When you are ready for mortar, set the stones aside in the same position and start laying them either in the same order, or a new order (FIG. 11-4). These are the same stones in the same position as seen in FIG. 11-3, but they are put in using a different sequence. They are started in the corners, brought down the sides a little, then filled in the middle.

When you are actually laying the stones in the mortar, it would probably be messier to lay them in the order you discovered them in. If the room getting the floor has two accesses, you could do the floor one-half at a time. You could leave half of the dry-laid pieces where they were until the first half is done. This way will be less confusing and take less space when you lay the stones aside.

There are other reasonable approaches to laying flagstone inside. As long as you don't have to walk on the new work as you work, and as long as you don't have to waste stone and cutting, it should be okay.

Drainage

Drainage on a masonry floor is important. Water should run off stones and not collect in puddles. Even an indoor floor needs drainage. On a below-grade inside floor, drainage is a nice flood precaution, and it makes cleaning a floor easier. If you want your flagstone floor to drain, it will need more slope than concrete. The rougher stone will create all kinds of tiny lakes. With a decent slope, they will be minimal in size. For a patio, 4 inches in 20 feet is not too slanted, but it should keep the water off. Determine the slope by the guide boards, lines, or stakes when the concrete base is poured. By knowing level, you can find 4 inches off level in 20 feet using a tight string, a good level, and a measuring tape, or using a water level or transit.

Having a slope on an interior masonry floor also means you need a drain at the lowest place. You can minimize the overall effect that a slope and drain will have on a floor by having the drain in the center. To get the equivalent of 4 inches in 20 feet slope, the center would need to be 1 inch below level and the ends would need to be 1 inch above level. Sloping

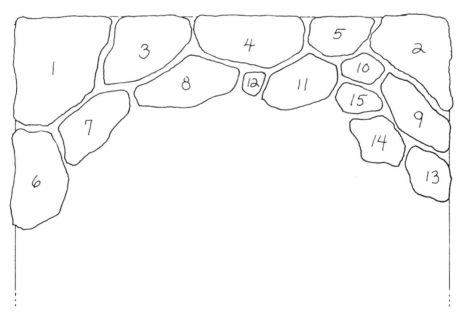

Fig. 11-4. The same stones can be installed in a different order once the arrangement is established.

toward a midpoint allows the entire perimeter to be at the same height, which makes things easier to plan and build. A slab that sloped from one end to another, on the other hand, would result in walls of varying heights.

You can lay drainpipe in the gravel bed under the slab, or you could build it into the slab. The opening for the drain must break the surface of the slab, and the other end must gradually slope to a drain field away from the house. Leave extra pipe sticking up when you pour the slab so the drain doesn't get clogged. You can cut it off at the finished floor height later. You can either bring stones around the drain hole or use a single stone to encompass the drain hole.

If you establish a plumb bob from the ceiling that marks the hole in the floor, you can cover up the hole without losing track of where it is. The drainpipe gets cut off at the top of the slab. Find a big rock that fits where the drain is and covers the hole. Mark the position of the hole (using the plumb bob) on the stone, take the stone away, and drill a hole at that spot. Put the stone back in to make sure it fits, then put a dowel rod or other temporary filler in the drain hole. This will keep mortar from going in the drain when you lay that stone in mortar. The stone needs to be lowered over the drain-hole plug. When the mortar is partially set up, pull out the dowel rod and, using your fingertip or a little trowel, smooth the mortar around the hole and make sure all the spaces are filled under the stone in that area. When the mortar is very hard, you can cut a bowl-shaped depression with a stone chisel around the hole to facilitate drainage.

FIGURE 11-5 shows a floor with a drain done this way. The drain is just below the top of the metal wood rack. Notice that it doesn't stand out like a big white plastic hole with all the stones circling it to bring even more attention. This flagstone floor is in a greenhouse, and gets frequent hosings for the plants, so it was essential that the floor drain. The stones are random brown sandstone, tightly fit with a dark mortar. This mortar gives a subdued feeling. The stones have layers that add to the beauty, but would weather poorly outside.

Outside Drainage

The patio shown in FIG. 11-6 drains slightly to the lawn, and the lawn slopes away from the house. This

Fig. 11-5. This irregular flagstone is laid with small joints of dark mortar. There is a drain hole in this floor.

situation coupled with the small watershed of the patio, required no further drainage. Incidentally, I built this patio 10 years ago, in limy mortar, and it hasn't cracked yet. The white mortar has gotten stained in places where metal objects were left to sit for a year. I could have avoided the stain problem by using the dark gray mortar as in FIG. 11-5. I got that dirtier color by adding a shovelful of black cement to a regular batch.

In FIG. 11-6, notice the larger perimeter stones and the few small fill stones. In the foreground, there is a stone that has only a point on the perimeter. This is a good way to avoid using a small piece to fill a small part on the perimeter.

Suppose you are making an outside patio adjacent to, and continuous with, the back of your house (FIG. 11-6). You need to slope the patio away from the

Fig. 11-6. A patio needs to slope away from the house.

should be about right. It will be softer than the stone, but able to withstand the abuse of people walking on it. Very thin joints (less than 3/8 inch in width) require fine sand in a richer mix.

The pointing mortar should be very pliable, but not smeary. Working off a trowel or a hawk, push the mud with a pointing trowel into the spaces till it is flush (and slightly more) with the edges of the stones. Try not to spill the mortar. It is a hassle to clean a floor, especially indoors. On a wall, you can knock the excess off and it falls to the ground. On a floor, any excess mortar must be dragged across the floor and swept up.

Point as much as you can with a small batch of pointing mortar, then go back and dress the joint as you would on a wall. Scrape away the extra mortar and bring down the surface to a uniform texture. You need to fill the joints a little bit too much so you'll have some slack when you dress the joint.

To dress the joint, the surface can be *struck* with a pointing tool, *stoned* with a stoning tool, or brushed with a small, stiff paintbrush. All of these procedures bring a little cement paste to the surface of the wall. Just touching mortar will bring moisture to its surface. They also tighten up the joint. I prefer the brush for the final mortar fussing because it is easier to get a continuous look. There are no trowel marks to cover or black marks from the stoning metal, and the brush knocks away protruding sand grains, while the harder finish tools push these grains back into the mix at the surface. Removing the protruding sand leaves a harder mortar at the surface of the joint.

You will probably have quite a buildup of mortar scrapings on your new flagstone floor when you're done pointing and dressing. Don't move them any more than necessary to see the mortar joint you are scraping at. If you try to clean up or sweep the scrap-

house so water won't move toward the siding. However, suppose your lawn or yard sloped toward the patio (FIG. 11-7). This would create a drainage problem, because water coming off the patio would build up at the bottom, having nowhere else to go. Most of the water in a light rain would percolate through the soil, but once the ground is drenched or frozen, the water will make a big puddle. To alleviate the problem, lay a drain along the bottom edge of the patio. The perforated drainpipe should be long enough to carry water to a lower level. During a long rain, you should see water flowing from the end of the drainpipe, but no puddles at the patio.

POINTING AND
FINISHING FLAGSTONE

When your stone is all laid in strong mortar and set, clean it with a hose and broom to get all the dirt and cement drips off. While the stone is still wet, but not with puddles, start to point the joints between stones. A mix of one part masonry cement to three parts sand

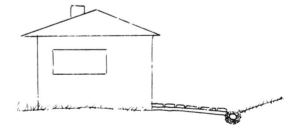

Fig. 11-7. A drain must be provided if the yard slopes toward the patio.

ings while they are fresh, they will stain everything they are swept across. Let the drips and the scrapings of mortar sit until they harden. Then sweep them into a dust pan in frequent small piles so you don't have to sweep the stuff far. The following day, use a wire brush or very stiff bristle brush and water to remove any stains. It is nice to be able to hose the mess off a masonry floor. Inside a house, that isn't always possible.

Curing the masonry floor is important, especially if it is outside in the sunshine. It should be covered and damp for several days. This is more important with floors than walls because floors take more abuse and need sounder mortar, and the large surface area of a floor is prone to drying quicker than a thick wall.

COATINGS ON FLOORS

I don't like coatings on stone. The best thing about stone is its imperious surface, its local, unique, and nontoxic nature. It is something of a trend to cover stone (and brick) with one type of plastic coating or another. There are several products on the market for coating stone. If you keep your plastic covers on your lamps for years and have plastic covers on your furniture, you might like the effect of plastic-coated stones. They will keep the stone bright and shiny and allow you to clean stains. I like things to wear and age elegantly, including lampshades, couches, and stone floors. Wear gives character. If your stone floor gets dirty, clean it with water and a brush, like the old days. If you feel compelled to coat the stone, use wax.

LAYING FLAGSTONE VERTICALLY

FIGURE 11-8 shows a house veneered in flagstone. I don't know if this is supposed to make the structure look like a stone house. It doesn't. Stone used this way serves no structural purpose at all, nor is it great siding material. It is prone to falling off, and all the stone is laid with the bedding plane in the wrong direction.

If the stone siding doesn't serve a real function, then it is strictly for looks. The trouble there is that it doesn't look like much either. This building could have been sided with cedar boards for less money, and it would have been much more legitimate, architecturally speaking.

Fig. 11-8. A veneer of flags does not look like a stone house.

The stone here is a hoax. In a way, it fits with the other hoaxes in this type of modern construction. In the same photograph, notice the nonfunctional plastic shutters for the windows? These are reminders of a bygone era. The siding on the roof overhang is aluminum made to look like wooden siding. On the roof are asphalt shingles made to look like slate. The idea of one material pretending to be another is so common these days, nobody really notices. I hate to see stonemasonry dragged into this charade.

If you decide to make a wall that looks like a stone wall, it definitely helps to know what a stone wall looks like. Stone walls are generally built of flattish stones laid on the broadest face, stacked in courses, with staggered vertical spaces. The corners of the wall are

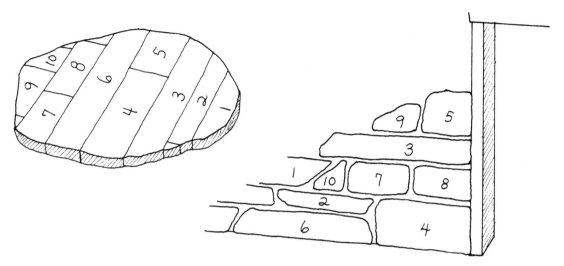

Fig. 11-9. A large flag can be cut into convincing shapes for a facade.

kept above the middle until the top. The top of the wall is capped with larger stones. Large stones would also be likely on the bottom course. To create this illusion with flagstone, keep the real thing in mind.

Suppose you wanted to stack a 2-inch facade against a block wall. First, you need to be sure there is enough footing to carry the weight of the stone; you can't stack a facade on the dirt. Even though it is so thin, it still needs a legitimate footing. Next, you must have some physical connector between block wall and stone facade. Normally, this is done with little strips of waffled sheet metal. Let the ties protrude from the mortar between occasional blocks, and bend them as needed to slip into the joint between stones. The waffles or corrugations in the strip of metal help the mortar get a grip. You should have one of these connectors every 4 square feet or so.

Next, make sure that the ends of your walls are not exposed. If they are, your facade will be revealed at the ends. Inside corners do not reveal the ends, and they are best for a convincing facade job. If you have outside corners (FIG. 11-8), you could bring the stone up to a massive corner post of wood to hide the thickness of the stone. Similarly, at doorways and windowsills the edges of the stone should be hidden behind wood. If you really wanted a place to look like a stone house, you would need to make the windowsills very deep, as they are in a thick-walled stone house. It amazes me when people think they have a

stone house and the walls are only a few inches thick.

The most important step in imitating a stone house with flagstone is to cut the stone into long pieces that can be laid in some semblance of courses. Make the shapes realistic representations of stone shapes used in a laid stone house and stack them in a similar sequence. Build up the corners first, with a long and large stone, then bring up the middle, crossing the joints as you go. Make certain the stone is square edged so it will stack on flat ledges. Outsloping tops when you only have 2 inches would be disastrous. If you cut the flagstone with a saw to get your realistic pieces, make sure the cuts are square edged. FIGURE 11-9 shows how a big slab of flagstone might be cut to yield pieces more closely resembling shapes that could stack with stability.

The mortar mix for a facade must be strong and sticky. One part masonry cement and two parts sand should do, or make a regular 1:3 mix and add a little extra portland cement. There will be no space to keep the mortar recessed, so you must do the pointing as you lay the stone. The mortar must be pretty runny to get behind some of the spaces it must fill so you will need to contend with some mess.

Mortar messes can be cleaned with a hose and a wire brush the following morning. If you wait longer, you will need muriatic acid to clean the stones. It is sold in hardware stores. Read the instructions on the bottle about dilution before using.

187

REPAIRING & REPOINTING STONEMASONRY

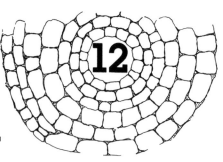

12

THIS chapter will deal with making repairs on existing stonemasonry. I will take you through two actual, and perhaps typical, repair jobs: the repair of a blown-out retaining wall, and the rebuilding of a stone wall that had partially collapsed because of the failure of wooden lintels.

I would like to emphasize the value of repairing stonework. Most people don't realize the amount of labor represented by a broken stone structure. All those stones had to be gathered to the site. If the structure is old, chances are some of the stones were dressed with a hammer and chisel. It is a pity to throw this effort away if it can be saved with relatively little effort.

One of the ironies of deciding to destroy an old stone structure is that it isn't easy to get rid of one. It won't burn away like an old wood barn. It is hard to push over old stone walls, and even if you do, there's still a big pile of stones to deal with. Sometimes, it might be about as easy to fix the old stone building as throw it away.

If most of the structure is still in good shape, repairs could make the building last for centuries longer. In some cases, a stone ruin has historical value that should be preserved. Some stone structures that now have no historical significance are certainly in a better position to one day have historical value. If you build something brand new in good stonemasonry, it might well be around 500 years from now, brimming over with historical importance.

FIXING A BLOWN-OUT RETAINING WALL

One of the most common repair jobs in stonemasonry is fixing ruptured retaining walls. The blowouts usually

occur from one of the following reasons:

☐ The wall is poorly built of round stones and eventually falls apart.

☐ The wall is too thin and doesn't have enough mass or integrity to withstand the push of the earth behind it.

☐ The wall has too many hollow places and is too flexible.

☐ The wall has no foundation, and eventually sinks into the dirt.

☐ The wall is infested with roots and vines, which pull it apart.

☐ The wall is entrapping water that is freezing internally and pushing stones apart.

☐ The wall is holding water behind it, and the hydrostatic pressure increases.

☐ The wall was actually built very well, but conditions have changed, and more force is now being placed on the wall than it was designed to handle.

The wall shown in FIG. 12-1 is one sidewall of an old stone canal that still drains the town of Gettysburg, Pennsylvania during high water. A large section of the wall collapsed into the canal and I was able to win the bid by being less expensive than the concrete contractors. I mention this to show that on some jobs, stonemasonry is competitive with other types of masonry.

The reason for the failure of this wall is a combination of all the above reasons, although I believe the final blow was dealt by a lovely mulberry tree. The tree, which was 8 inches in diameter, was growing behind the wall at the edge of an alley. It had roots as big as arms all through the wall, right to the bottom. A hard wind blowing the tops of a leafy tree could add some major stress to the wall gripped by the roots.

This wall was also a bit on the thin side for all it had to do. The area that collapsed had a distinct shortage of *bondstones*, or stones that go deep into the wall. You can see from the photograph that the wall has two distinct vertical layers. If it had had three layers, or two that were tied together well, the wall wouldn't have failed, even with the tree.

This wall also had a water problem. An alley and several houses were draining their rain to a spot behind the wall with no access for it in the wall. All that water had to run down behind the stone wall. Another big problem is that, when the wall was built, it had only to retain the earth behind it. Now there is a road on top of that dirt behind the wall. The alley nearby was used for horse-drawn carriages and very few of them, because there were no houses nearby. Suddenly, there were a lot more houses, and none of them soak up rain water, so the natural drainage is stressed. When it rains now, water flows down the alley way, because it is lined with houses that drain rain that way. When the wall was made, most of the rain was soaked up by the soil. One hundred years ago, it would have been hard to imagine that the wall you were building would someday have to support the weight of a fully loaded concrete truck that made a wrong turn down an alley.

I recently repaired a stone bridge that had faithfully served a back road for 150 years. It was destroyed by a huge fire engine that was out on a fund-raising drive and chose the wrong driveway. You can't say the thing was poorly built; it was loaded beyond its intended capacity. Even a modern bridge would be destroyed by bumper-to-bumper concrete trucks. I've repaired a number of retaining walls that were built to hold back dirt, but they ended up supporting a parking lot on top.

Some weight beyond the weight of the dirt is okay, and even a car isn't too much for a decent wall. The problem often starts when a very heavy vehicle backs up on the little parking lot on the ledge that is supported by the retaining wall. The weight is too much and the wall buckles.

Remove the Damage

There is no easy way to fix a wall that has collapsed, especially when it had been holding back earth. You have to get in there and pull all the rock out of the way. As you take stones away, dirt will spill down to take its place. This is very frustrating if the dirt is loose and it just keeps spilling down, ton after ton. It all needs to be removed and eventually put up behind the wall again. Sometimes this kind of work is best done when the ground is frozen and the dirt won't collapse behind the wall. The wall in FIG. 12-1 was repaired in summer, to avoid standing in water while we worked. Unfortunately, the stones that had fallen out of the canal wall had been removed before I started.

The man in FIG. 12-1 has hose and trowel in hand. He is removing the ledges of dirt that remain on the stones in the rear. There was no mortar in this wall except at the surface, and that had been applied as an afterthought. Because I would rebuild the wall with mortar, all the dirt and dust needed to be removed.

Notice the big, cut, black-granite (diabase) capstones on this wall. The one above the damage is held in place because it is jammed between the caps on either side. When doing a repair job of this nature, there is always the danger of more stones falling at you. We gingerly knocked out all the stones that were teetering so they wouldn't hit us later. I was lucky that the back half of the wall held. We didn't have to move tons of backfill dirt. Including gathering 3 tons of stones, 1 ton of sand and five bags of cement, this job required 40 hours labor.

Restack the Damaged Area

After getting the inside of the wall clean and all the dirt off the footing, we started laying large, clean stones at the bottom of the wall. Where the patch job meets the existing wall on the sides, I tore out enough stone to ensure a good tie-in. More important in this case are stones that tie into the depth of the wall. When I found a long stone, I would create a pocket to connect with the back half of the wall.

The wall was rebuilt in courses, with overlapping joints. As far as appearances go, the wall I was building into was such a mess, I made no attempt to match its style, which was a hodgepodge of patch jobs.

Mostly for convenience, I used a mortar mix of one part masonry cement and three parts sand for the whole job. The mortar was left recessed as far as possible from the face of the wall, but it still managed to drip all over the stones. On this job, it was more

Fig. 12-1. *A variety of forces caused the failure of this retaining wall.*

Fig. 12-2. *The big stone will give this wall greater stability lying flat.*

Fig. 12-3. *In a break with the normal order, the back side of the wall is built up first to stabilize this shaky situation.*

important to have the wetter mortar key-in to the stones in the back of the wall than to have a dryer, tidy mix.

FIGURE 12-2 shows the big vacancy in the wall where the disturbed stones were removed. This is a tenuous situation. If you need to do this kind of repair, stay ready to step lively. There is a natural arch effect taking place in the back of the wall, but it is marginal. For this reason, I broke with the normal logic in stonemasonry, and brought the inside, unseen side of the wall up higher than the face. This procedure is evident in FIGS. 12-2 AND 12-3. The vacancy behind

the big stone in front was filled with stacked stones, much like the ones that were already there from the Civil War days.

In FIG. 12-3, you can see the new stones laid in. I smeared the faces with mortar for added stability when the space was filled.

I was still nervous about the 800-pound capstone until the rest of the back was filled. I took advantage of the missing dirt behind the wall by letting some stones reach back into the bank. Stones tied in that way cannot tip over because of the weight of soil above them. It is like holding up a temporary high-

Fig. 12-4. The wall is repaired, but not pointed. Drainage has been added.

Fig. 12-5. The pointing is finished and the wall is sounder than it ever was.

way sign with a bag of sand on its base. It is good to do this only below frost line so that anchor stones won't be pushed upward.

I made the wall stronger as I rebuilt it. The addition of mortar not only adds adhesive strength to the wall, it adds a lot of mass. There were many empty spaces in this wall that are now filled with heavy material. The wall is heavier. Small stones were replaced with larger ones. Merely putting the wall back like it was would probably not be enough, because it fell over once. By continually creating level surfaces, we brought the wall up with stability.

Because there are no real ends or corners in this repair, I didn't need to keep the ends higher than the middle. Instead, I concentrated on the center of the rupture, and got the big, deep stones in there by making one of them the first stone of each course. I would lay the big stones over a series of smaller ones below to get the most joint crossing with the rocks. If all you have are big, long stones, joint crossing will be easy. If you are short on the big ones, put them where they will do the most good.

FIGURE 12-4 shows the wall all filled in. The mortar has been scraped back, and brushed and hosed off the wall. Near the top of the wall is a plastic drainpipe. This pipe was sent through the wall and is exposed on the other side to accept runoff from roof drains that had previously been draining along the tire treads of the alleyway. This water had been seeping through the wall, adding to its woes. The pipe was left protruding from the face of the wall to keep the water from dribbling down the wall. It doesn't look too great, but the stone wall is in a canal, and didn't warrant anything fancy. Notice the pipe passes through the wall in an area that needs no support. You can't expect a plastic pipe to be one of your stones. Create the space, then put the pipe in.

The wall is ready to be pointed in FIG. 12-4. It took 28 hours (14 hours for two men) to get the job to that point, not counting the gathering. It is tricky to time stonework, and this makes it complex to have helpers. My helper on this had only one day he could work, but he said he'd work from sunup to sundown—and we did—though we paid in backache the next day. I came back alone the next day to point the wall, which took another 8 hours. On a job this size, with about 50 square feet of face, there is not enough room to

have two people pointing without getting mortar in each other's hair.

Knowing roughly how much you can point or lay in a day will help you guess how long a job will take. If this pointing job had been much larger, I probably would have pushed myself, rather than go through the bother of setting up the next day for a short job. I don't think you could point 100 square feet of rubble stone in a day if you were doing the whole job (mixing, etc.) without it being pretty sloppy. Although I'm sure there is someone out there who can point 200 square feet of stone in a day, try for 50 square feet, and do it in two batches.

With a small batch of the 1:3 mortar mix (9 small shovels of sand and 3 of cement), I could point half of the wall. While this first batch was getting firm, I mixed another one like it and filled in the bottom half of the wall. It was a bit ridiculous to make this pointing look real great, in view of the patchwork on both sides and all along the canal. I moved faster than on the inside of a house.

By the time I finished putting the mortar in the cracks on the bottom half, the joints in the upper half had set enough to scrape down without smearing. Digging back the mortar with the tip of a pointing trowel and knocking off the stray blobs of mortar takes about half as long as putting the stuff in. When the top half was dressed down, I gave the joints a brisk brushing with a small parts brush.

With the top finished, I checked the consistency of the lower, undressed mortar. It was still too soft. It is worth noting that pointing near the ground takes much longer to set up than pointing higher up. It is cooler and more humid nearer the ground. Mortar made during the heat of the day will set quicker than morning mortar when the ingredients are cool. For a strong finished product, don't let the sun shine on the mix. It will heat it and dry it. In the summer, I keep a scrap of foam board to cover the wheelbarrow. During this job in July, the wall needed frequent mistings from a garden hose.

Because the lower mortar was not dry enough to clean, I knew it was time for a long lunch break. Breaks during masonry tend to be dictated by the batches of mortar. After a leisurely lunch, the mud was stiff enough to finish dressing. The next morning, after arranging to do other things in the area, I gave the wall a quick going over with hose and wire brush. This removes last-minute smears and gives the wall some more curing moisture. The process only takes about 10 minutes. I wouldn't have been able to do this final cleaning on the same day as the pointing because the mortar would come right out. If I had waited two days, the mortar wouldn't come off very easily with hose and brush. This 10 minutes with hose and brush is the most effective minutes you can spend on a stonemasonry job for its appearance.

FIGURE 12-5 shows the wall all pointed, dressed down, brushed, and hose and brushed. The old stone wall is better than it was 100 years ago, and it cost the borough less than a concrete repair.

REPAIRING A COLLAPSED WALL

Old stone buildings and foundations that appear to be ruined might well be worth restoring. Most of the work that went into the original structure was in the shaping and hauling of stones. In the case of the old

barn foundation in FIG. 12-6, which is to be repaired to sport a house on top, we're talking about 135 tons of carefully selected fieldstone. That much of the job is done—an incredible head start even if all the stones need to be relaid. In the case of this pre-Civil War barn foundation, 100 tons of the stone were still in the right position, needing only repointing. Yet at first glance, the thing appears to deserve a good bulldozing.

Bulldozing is frequently the fate of stone leftovers in this condition. Structures of this sort have lasted 800 years. You must weigh this consideration when deciding if a relatively small effort is to be made to save the structure for future generations. Two-foot-thick stone walls means playing for keeps in the construction world. Getting your hands into a project of this sort gives you a feeling of connection with the past and creates an affirmation of the future.

On this repair job, the main problem areas were the result of wooden lintels that rotted, allowing the stone above to collapse. Also, the top plate timbers

Fig. 12-6. Rotten wooden lintels let this old structure start to crumble.

along the stone walls had so thoroughly decayed, that the humus was supporting a wide variety of vegetation, the roots of which had caused considerable damage to the top course of stones.

The window opening in FIG. 12-6 is what we had after cleaning stone down to the level of the busted wooden lintel. The dirt was cleaned off, and a level, secure pad was constructed on either side of the window opening to accommodate the stone lintels that would replace the rotted wooden ones and allow us to build the wall back up. The owner of this project had the dubious fortune of acquiring a pair of 1,200-pound granite blocks, which were to be the new lintels.

Bedding Mortar

All the stones that needed to be relaid were set in a weak bedding mortar, to be pointed later with a richer mix. To ensure space for the 1-inch-thick pointing mortar, the bedding mix was scraped back crudely before it set.

There are several reasons to bed stones and point the walls as separate procedures. For one thing, you avoid ever working above the finished product; you finish your way down the wall without making a mess. By laying up and pointing down, you also avoid shocking the finished joint with the big stones being set above. It is much less crucial when the bedding mud gets jarred and cracked from new work, as ours did when the granite lintels were lowered into place.

A bedding mix can be made very weak and still be quite adequate for the job of a 2-foot-thick wall even three stories tall. The mortar used to bed the stones in this old barn foundation is indistinguishable from dirt. Indeed, there were some healthy plants growing in the stuff. I have encountered the same phenomena in many old stone buildings. You take one apart and you have a pile of rocks and a heap of dirt, about one-third as large as the stone pile. The clayish bedding mud is kept from eroding out of the wall by a pointing joint of lime and sand.

Modern masons and builders will cringe, but the bedding mortar in a laid stone wall needn't be very sticky, nor have much compressive strength. A typical cubic foot of stonework weighs about 144 pounds, exerting a force of about 1 psi on the mortar below. At the bottom of an 8-foot wall, the force is up to 8 psi. A two-story building on top of that might double

the weight. There is simply no advantage in using a mortar mix that can support thousands of pounds per square inch when all you need to support is 100 psi at most. Actually, there is a distinct disadvantage in using too strong a bedding mix.

Factors Affecting Mortar Setting

Masonry in general, and mortar specifically, is an ongoing experiment. Methods and mixes vary like the wind, and time is the best critic. We do know that hard mortars, with too much portland cement, can cause serious problems in a stone or brick wall. Portland-rich mortars contract on setting, causing cracks and opening waterways in the wall. Also, a mortar as hard as the masonry will not act as a shock sacrifice.

The walls that last the longest do so because they can handle a certain amount of stress and motion. The mortar will give, sparing the stones. This is significant in offsetting the effects of expansion and contraction. In most ancient walls, each stone, in effect, is wrapped in a relatively insulating blanket.

More subtle, but significant, is the way that soft mortars breathe and allow the passage of humidity. Experience with waterproof coatings is building a body of evidence that allowing no passage of water vapor can cause spalling on the surface of the masonry. I don't think absolute waterproofing is a good idea in masonry or wood buildings. Water needs to be kept out, and a good pointing job will do that, but humidity should flow more freely.

Mortar mixes behave predictably, but the factors affecting the way they act are many and variable. There is no final word on cement mixes for particular jobs. It helps to pay attention to each unique situation and to have a basic working understanding of the chemicals and reactions you deal with in masonry. As shown in FIG. 12-7, sand with grains all the same size has larger spaces between particles than sand with grains of varying sizes. It requires more cement and water to fill those spaces.

Round sand, often cursed in masonry books as worthless, has been used in mortars that last centuries. Round sand has larger spaces than square sand, but because round sand nearly always has a wide variety of pieces, it might have less spaces than sharp sand. Round sand flows better than sharp and needs less cement to achieve workability. Manuals for masons abound in contradictions, although most

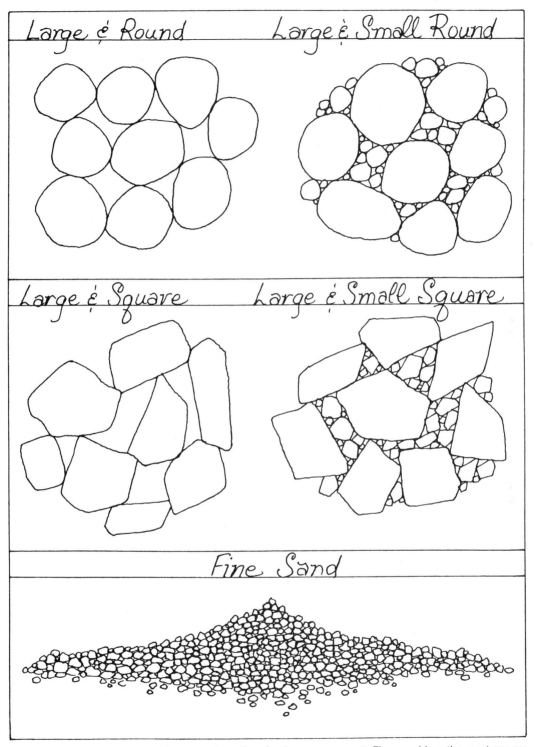

Fig. 12-7. Sand has less spaces when a variety of grain sizes are present. Fine sand has the most spaces.

agree that varying sized sand grains will best stretch your cement dollar, giving you the strongest mortar for the money.

Most sources agree on using only drinkable water in your mortar mix. It is true that the optimum strength mortar is made with pure water, but if all you have is grossly polluted water, you can still whip up a suitable concrete for a footing or a bedding mix for stonework. In fact, you can make concrete with sea water that would do for most stuff around a home. Don't be afraid to get a few bugs in the bedding mud. It is not that fussy a thing.

The bedding mortar used in this project was by volume: 1 part portland cement, 2 parts hydrated lime, and 13 parts regular mason's sand. Water was added gradually to achieve a workable paste. TABLE 12-1 shows some other possible mortar mixes. Water quantity can vary a lot, depending on how wet the sand is. The water quantity will affect the type of product you get more than the other factors. Generally, the least amount of water it takes to get a mixture that is uniform and pliable will give the strongest mortar. If you need wetter mortar, add extra cement.

Temperature of ingredients plays a significant role in the way mortar acts. It is best to keep all the components between 40° and 80°F, although 50°F to 70°F is better. Colder temperatures generally cause slower setting; warmer, faster setting. In the summer, it is easy to get fooled and make a batch of mortar with a hose that has 140°F water in it from the sun on the hose. If the sand is hot, too, the stuff will stiffen before you can finish mixing.

The biggest plus of using a separate bedding mix is that variables in the weather and other goofs have less chance of wrecking the job or halting production. If the bedding mix gets a bit too hot or cold, or wet or dry, or if you lose track of the amount of cement on one batch, it is of little consequence. Whatever you come up with has got to be better than the clay, lime, and dirt between the stones of many historical stone buildings.

It is the pointing mortar that should be just so, with the right conditions. If your pointing mortar gets too hot and sets poorly, it will come out of the wall in the too near future and allow the crummy bedding mud to begin eroding. If a pointing job freezes before it sets, you might as well tear it out.

Repairing the Wall

After we prepared the space for the lintel stones, we managed to entice some passersby into being a crew for an hour. They demanded to see the cold beer first, but we soon managed to build some crude ramps. This being a construction site, there were plenty of 2 × 10s and cement blocks around. This fact played an important role in the decision to employ such primitive technology.

I have learned it is important to not lay the ramp right on the wall when you are using a bruiser like this lintel stone. The top stones can flip up from the angle of force. The planks also get trapped under the rock if they extend beyond the face plane of the wall. Notice in FIG. 12-8 that the ramps come to the level of the prepared surfaces. With the stone one flip from the top, we adjusted its position on the ramp so that the show face would end up at the right spot in the wall.

FIGURE 12-8 has captured the moment of tension. For an instant, the stone is weightless. These boards are pinned to the ground at the bottom so they can't kick back. The boards are doubled and supported every few feet with a block pillar. Make all the last-minute

Table 12-1. Possible Mortar Mixes for Coursed Rubble Stonemasonry.

Possible mortar mixes for coursed rubble stone masonry

PORTLAND CEMENT TYPE I 94 LBS CU. FT.	HYDRATED LIME TYPE S 40 LBS CU. FT.	MASONS SAND GRADED	
		Bedding	
1	1	9	Harder
1	2	13	
1	3	15	Softer
		Pointing	
1	1	6	Harder
1	2	8	
1	3	12	Softer
		Parging	
1	1	4	Harder
1	2	6	
1	3	8	Softer

adjustments on the ramp. The stone will be much harder to move on the wall.

The bedded stone underneath the mammoth stones was busted up pretty badly when the lintel was lowered onto the wall. Using a car jack, set on blocks on the window sill, we pumped the stone up a few inches and repaired the damage. We also pushed in a fresh bed of mortar with a bit of extra cement.

Lintels and arches often carry more weight than the other parts of a stone wall. FIGURE 12-9 shows the jack in use and the big stone floating. Its position is easy to fine-tune on this pivot. The photograph also shows the other end of the stone up off the wall while the man adjusts the rocks that were upset beneath.

With new lintels in place the wall could be built back to its original height. Strings were pulled from the corners to find the top and face of the wall. Scaffolding was set so we could work at that height. The corners came up first, then the middle was filled to that level. Similarly, within the thickness of the wall, one face was built up, then the other was brought to that height, and then the middle was filled in.

The overlapping in the center of a wall is important. You should avoid building a wall that could cleave in two. To achieve continual intersecting, one face must be brought to the height of the other. If you put a deep stone on one side of the wall and a shallow complement to the same height on the other side, on the next course, reverse the procedure to cross that internal joint. To be able to do this, you need to keep bringing stones to a new, level beginning. Uncoursed stonework has less opportunity for this internal intersecting of the separate sides.

FIGURE 12-10 shows smears of bedding mortar being removed with wire brush and hose. The limy mix should be hosed all the way down the wall or it will leave a chalky stain on the stones. At this point in the project, the lintel stones were carrying weight and most of the wall was back to its proper height. The parallel lines can be seen above the unfinished part of the wall.

Because this wall (and many others) is 2 feet thick, a 2-foot level becomes a very handy tool for setting face stones. It is the right length to use as a measuring stick. A 2-foot framing square is also handy. The top of this wall could be checked easily by sighting across the two top strings. No stones should penetrate the top line. If you plan to top a wall with a wooden plate of some sort, don't let a nubbin of stone cause you fits. It is much easier to add a little mortar at the top than to subtract stone.

Pointing the Stonework

Another advantage of using weak bedding mix is that it cleans easily and allows you to lay stones faster and sloppier. I usually spend about ten minutes at the beginning of the work day cleaning off the mess from the previous day's stone stacking. This cleaning with hose and brush also ensures that the wall will get a modicum of curing water.

When all the rocks were laid, the bedding mortar raked back, and the wall cleaned, we were ready to point the wall. Pointing mortar, the continual bead of permanent caulk that keeps the bedding mix from eroding out of a wall, must be harder and stickier than the bedding mix to endure the weather. The mix we used on this restoration for pointing was, by volume, 1 part portland cement, 2 parts hydrated lime, and 9 parts mason's sand.

We used white portland cement because it gives a whiter joint, which is more reminiscent of the old mortars and shows off the color of the stones better than gray mortar. The white portland also gives a mortar that is less staining and easier to clean. White is the easiest color to match when pointing. It is hard to see the dead joints where you stop a day's pointing with white mortar. White portland gives a slightly weaker mix, but in the case of laid stone walls, it is irrelevant.

A limy pointing mix, as the one suggested here, will expand slightly as it sets, giving a very tight seal around the stones. The lime will continue to seal its own hairline fissures as the carbon dioxide in the atmosphere reacts with the calcium carbonate in the mortar. This is probably why old lime pointing in parts of Scotland has endured 300 years, while modern pointing mix can't hope for that kind of longevity. The enormous hassle people used to go through getting lime from limestone is impossible to duplicate on a job site, so you can't really copy the old formulas that have passed the test of time. You can mimic them with modern materials, however.

Other factors being the same, mortar acts differently in different sized pieces. Thin crusts of mortar

Fig. 12-8. One flip from the top, the stone is adjusted to come out in the right position.

Fig. 12-9. Stones that were upset by the intrusion are adjusted.

Fig. 12-10. Weak bedding mortar mess can be brushed off easily the next day.

Fig. 12-11. The rough pointing is dressed with a small trowel when it has set slightly.

dry out very quickly. Fat globs of mortar cure more slowly. Very thick masses of mortar generate enough heat to set prematurely.

REPOINTING STONEMASONRY

Starting at the top of the wall, push the pointing mortar into the clean, moist joints between stones. Use a trowel or hawk to hold a blob of mortar and push it with a thinner trowel that barely fits the spaces between rocks. The mortar should be the consistency of peanut butter to get a good grab. When the void is full, move horizontally across the wall without stopping to pretty up the mortar. Leave the mud crude at this stage. It will be much easier to dress later when it has lost that smeary quality and become crumbly.

The less you manipulate the mortar, the better the set will be. Just dragging a trowel across the mortar surface will start to separate the aggregate from the cement paste. *Tooling* a joint is when you purposely drag a metal tool across the wet mortar to create a hard, slick finish surface. What I don't like about that approach is that if you break that hard crust, there is a weaker zone just beneath because the cement

Fig. 12-12. The finished, cured product is revealed.

was drawn out of it to the surface. If you don't tool the joint, the mortar beneath the surface of the joint is the same strength.

Too much fiddling with the mortar causes the reaction to begin again. You can bash half-set mortar into a pliable form again, but it will be as much as 50 percent weaker when it finally sets. *Retempering* mortar, as it is called, will allow the mud to remain pliable for much longer. This is a favorite trick of pointing crews on big brick jobs. All the books say don't do it. All the masons do it.

Adding extra water to a batch that is getting too dry will dramatically alter its final strength. With bedding mortar, I occasionally allow myself this convenience, but it should be avoided when repointing. Especially if you are pointing stone alone, keep your batches small so you won't have mortar getting stiff faster than you can use it. Then you won't be tempted to remix the stuff. Your batch will be about the size of the sand. Nine shovels of sand is about all I can handle in one pointing session. Batches this small definitely don't warrant the hassle of a motorized mixer. You can do it real quickly in a wheelbarrow, and push it right to the wall.

I use an old wheelbarrow pan for mortar on scaffolding. I can shovel the contents of the wheelbarrow into the pan. This common trash item is quite useful in masonry. I try to keep a pan directly under the work area to catch most of the drippings of mortar. If they are reused right away, the spills are okay. If the droppings get dirty, I use them in the next batch of bedding mortar.

For beginners, the amount of mortar that falls off a trowel when pointing can be frustrating. It also takes awhile to get the feel of the tools. My best tip is this: assume a lot will spill. It is all right. Try to get a catch pan or board below you so you won't get upset by the loss of most of the mortar. If you drip mortar on the stones—and you will—leave it alone until later when it is dry and inert. Don't worry about how clumsy the tools feel. A first-time novice can create a topnotch job with horrible tools.

Refusing to be intimidated is half the battle of learning something new. Ignore almost all the pictures of tools you have ever seen in masonry books. Forget about the professionals that charge a lot of money. The beautiful barns that grace the United States were built by farmers, not carpenters. Rubble masonry is

in that special class of building materials that doesn't need architects to look great; and doesn't need professionals to assemble.

Dressing Down the Mortar

After a small batch of pointing mud is used up, check the spot where you began. If the mortar is crumbly, you can start to dress the joint, usually with the tip of a 1-inch-wide trowel (FIG. 12-11). Dressing also can be done with a stick. I know one mason who uses a spoon to dress mortar. In the illustration, it is obvious how much sharper and crisper the dressed joints are compared to the rough ones below the man's trowel.

If the mortar is still sticky and accumulates on the trowel, leave it alone awhile. Timing is everything on this phase of the job. On a warm, dry day, the mix should be ready to dress in an hour. On a cool, wet day, the morning mortar might not be ready to dress until afternoon. A mortar that starts out wetter will take longer to set. Porous stones can absorb moisture from the mortar, causing a weak, fast set. Direct sunshine, especially on black mortar, can speed the set and should be avoided. An increase in the percentage of portland cement will cause a faster set, and all lime mortars are very slow and need faithful curing. It is all predictable stuff, but the factors are many and varied, like the rain, the sun, and the temperature.

For me, the dressing down can take half as long as the installation of mortar. If you underestimate this time, you are likely to point for six hours, start dressing down the mortar, realize you aren't going to get home for dinner, and let it go until the next morning. This is usually a bad mistake because it is so hard to scrape the next day. The other approach to underestimating the dressing time is to rush through the job and get a messy looking finish that will last your whole lifetime. If you want to do an 8-hour day of pointing, don't do more than 5 hours of actual pointing, and you will have enough time to tidy it up.

Curing

The ideal with pointing mortar, or with any cement mixes including concrete, is to keep the original quantity of water in the mix until it has entered into the reaction and become stabilized. Mortar sets up, it doesn't (or shouldn't) dry. It is best if the mortar, as it is setting in the wall, is kept moist and between 80°F and 40°F for five days or so. For a month would be better, but not realistic. This means no freezing, no heat waves for days.

If it isn't 60°F and lightly raining the whole time your mortar is setting, you must create those conditions artificially. This is best accomplished by draping burlap or other cloth over the finished areas and keeping the material wet with an occasional hose spray.

Almost no masons today take this step. It is more important with the limier mortars. It is nearly impossible for a mason to return for several days after the job is done to ensure proper curing. It is more the job of the owner to keep it moist.

There are several schools of thought on finishing mortar joints. Some masons form a convex protruding joint with a special trowel; others use a metal stoning tool to create a hard, slick surface. I prefer a recessed joint, mostly for the sake of appearance, but also because it is easier. With a recessed, brush finish joint, the stones are the dominant feature to the eye, and the mortar joint is subdued.

To achieve the finished pointing look seen in FIG. 12-12, I scrape back the initial joint filling (when it is crumbly) with a trowel until it is uniform and smooth. Then I give it a brisk brushing with an auto parts brush, which tightens up the joint by removing protruding sand grains and by bringing a little extra cement and lime to the surface. I remove persistent and evasive smears the next morning with a water hose and wire brush. This final going over with water helps spot any missed spots in the pointing, while giving the mortar a good soaking for at least a token curing.

Most stonemasonry done in the United States these days is for aesthetics. People understandably prefer the look of stones to block or concrete. Because of this, it makes sense to go through the extra effort to have a clean finished product. If you have doubts about spending the extra hour in ten that is required to get a first-rate job, think how long people will be seeing it. Again, stone walls of this nature have lasted 800 years.

Fine Points about Repointing

When you are repointing a stone wall, there is a limit to how large a lump of mortar you can keep in a wall. A fist-sized blob will often slump back out. I usually break up these large spaces by shoving small, clean stones into the vacancy. Deep holes and large spaces might need to be filled in steps, as you might with joint compound on a hole in the plaster board, or with wood putty in a hole in a board.

By working in horizontal layers when pointing, you will avoid the look of a dead joint or a cold joint. If you complete a vertical section on one day, and then do another section adjacent to it the next day, you will be able to see the junction of the slightly different shades of work, as you would if you were painting a wall. If you can't complete a full horizontal row in one day, end your efforts in jagged steps for a less obvious juncture.

The way the finished surface looks is strongly affected by the way the joint is tooled. If you want a uniform appearance, the same mason should point whole visual units of a structure.

When repointing an old (pre-portland cement) stone building, it is important to get the stones clean before you apply the mortar because mortar won't grab a dusty rock very well. A hose is almost essential for this step. If your job site is away from plumbing, it might be worth renting a pressure tank of some sort, or hiring a water cleaning company to wash the walls.

If the gaps are big and the bedding mortar is dusty dirt, the washing can start a seemingly endless flow of dirt out of the wall. Don't let any more escape than is absolutely necessary to get the pointing surface dust free. Remember, the stones are sitting on the dirt bedding, and if you wash enough of it out, the wall will start to come apart. The deeper the stones are, the less a problem this is.

If you have very deep holes to point, you should do it in stages, although this method isn't convenient when you're setting and tearing down scaffolding. That's another one of those things that is by the book that nobody except a government employee would ever do. One manual on pointing I've seen says that if your joints are 1 inch deep, you should refill the space in two sessions of 1/2 inch each. If you can find someone that will pay you handsomely for this extra effort, it might be worth doing that way.

FIGURE 12-13 shows an old (150-year-old) stone barn near my home. It is losing its original lime and sand pointing mortar. This is to be expected, and it

Fig. 12-13. The original lime mortar is starting to fail in this old barn.

wall. This is easily done if you fill the big space completely with a fairly stiff mortar and then push the small, clean, moist stone into the mortar, making sure you keep a good space all around the small stone.

Buildings that haven't been repointed with portland cement mortar are a joy to repair because the lime mortar comes out so cleanly and easily. Do the next guy a favor: use a lime mortar when you repoint your stone or brick house. If all of it comes out white, it will be much better for future patches. You might have one section, in areas of harsh weather, that only lasts 50 years. You can't wait until the whole building needs to be repointed before tending to that area, so repoint a section. Shades of portland mortars vary considerably and old buildings get covered with multicolored patches, which is a pity for such treasures. White is white and is easy to match.

A good pointing job is a credit to a good foundation and well-laid walls. A pointing job cannot be expected to hold together a failing stone building, however. Don't be tempted to do a pointing job to try to save a wall that is failing as a result of an inadequate footing. It won't work, but people try it all the time. When they finally get someone to try to fix the real problem, they have to contend with a wall all glopped up with incredibly strong mortar. In most construction failures, check the drainage first. Then fix the footing. Then, if the stones are laid well, a pointing job might be in order.

Make sure you know what you are getting into when you start a pointing job. Often, a homeowner will decide to fill the spaces in a dry-laid wall. This is a waste of time and will often hurt the wall by preventing free passage of water and by trapping water inside the wall. It also will make a dry wall ugly and cause it to start to break apart at the first sign of normal motion.

Old house foundations are sometimes dry-laid, and the basements are wet. A common blunder is to try to stop the moisture by pointing the basement walls. The basement is wet mainly because there is no drainage system to carry the house's rainwater away from the walls. Without developing a drainage system, pointing the basement walls will only back up water and increase the hydrostatic pressure on the walls. If the water is back there, it will come through the wall somehow, no matter how much glop you smear on the walls.

isn't because the barn was poorly built. A stone building needs to be repointed every century or so—it's part of the deal. Using stronger mortars won't change that.

This barn is badly in need of repointing because the weather is washing out the clay bedding mud. When the bedding mud erodes, which doesn't take long because it is so soft, stones can fall out of the wall, especially smaller ones that don't go deeply into the wall. A tightly fit wall will not suffer as much erosion when the pointing fails.

Most repointing, since it is expensive work, is not done until it is badly needed. For this reason, most repointing involves putting small stones back in the

Fig. 12-14. The stone must be taken down to the level of the broken wood to repair this wall.

I have heard of people pointing old stone wells that were lined with dry-laid stone. First the thing goes dry, mostly from a steadily lowering water table, then someone gets in there with mortar and patches the holes. Because the dry walls are below the frost line and because the inside of a well is about ideal mortar setting conditions, the pointing holds and keeps the water from reentering the well when the water table comes back up. Be careful not to point weep holes in the bottom of a mortared wall. These holes are left in the bottom of a wall to allow water to pass through, and if you try to stop it, pressure will build.

If you attempt to repair a stone structure, first look for the possible causes of failure. If a wall is too thin and poorly made of round stones, it isn't worth saving with a pointing job. To save a wall like that, build a block wall or pour a concrete wall against it.

If you have a ruin that resembles the old barn foundation in FIG. 12-13, it is worth saving. FIGURE 12-14 shows a different place on that same structure. This is on an inside wall that had been whitewashed. Don't be fooled by whitewash, which makes old stone look so bad. It comes off pretty easily, and in this case, the wall is in better shape than it first appears. Notice the rotted beam in the stonework. That had to be removed, and all the stone torn down to that level. Old stonework is not glued together in the least. These rocks can be lifted off the top of the wall, one by one. This section of wall was rebuilt without wood. The lintels of wood that can be seen over the windows here are still sound, but if you are already doing repairs, you should consider replacing all the wood that is supporting any masonry. It will become a problem later, even if it isn't right now.

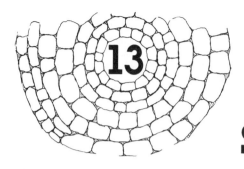

13 SPECIAL ASPECTS OF STONEMASONRY

IN its beginnings, masonry was synonymous with cutting stone. A *stonemason*, or mason, was someone who worked at a quarry, chiseling chunks of rough quarried stone into uniform building blocks for use in a preplanned structure. The men who assembled the stones at the job site were called *fitters*. The stonemasons sometimes didn't ever see the buildings they were helping to build. In a sense, stonemasonry was a factory job.

Things are different now, with the advent of portland cement, concrete, and metal-reinforced concrete. The scope of masonry has broadened into new areas, and stonemasonry has faded into near extinction, in its classical sense. I'm not nostalgic. Concrete is a great idea whose time had to come. There's no sense laying stones where you can't see them anymore, unless you can't afford concrete. Even with free stone on hand, it is hard to beat the price of concrete delivered to your site.

To build a true ashlar masonry house these days, you would need to be very wealthy, or have a near obsession with stonework or a particular period of history. In Buffalo, New York, there is a rather famous stonemason who is building a real stone house, and he's been at it for over ten years. Naturally, his place still looks like a construction site—it is. The city has been trying to get the place torn down because the construction site is an eyesore in a residential area. The irony of this is that the man's stonemasonry is first rate. The idea these days is to get your place up in three months and mow the lawn on weekends. Some of the big stone cathedrals in Europe were a mess for 150 years while they were being built.

ASHLAR MASONRY

Ashlar stone is any stone that has been given a particular shape by the hand of people or machinery. It normally refers to a relatively thin and square stone used as a facing for rougher masonry. Ashlar stone is designed to fit closely together, with mortar joints as thin as 1/8 inch. The stones themselves are often wrought by hand to tolerances of 1/16 inch.

The paradox of trying to obtain an ashlar look with uncut stone is that you're always looking for the square stones. If I was rock-hunting, and saw a chunk of sandstone shaped exactly like a cement block, I'd squeal for joy and snatch that thing up before some other stonemason found it. Yet, if all the stones were shaped like that, the wall would be boring to build and behold.

In the old days, the boredom imposed by cut stone was offset with incredibly elaborate doorways and cornice work with gargoyles and steeples. Today, that kind of detail is pretty much out of the question. A true rubble wall now has the possibility of being the loveliest of buildings—solid and handcrafted.

FIGURE 13-1 shows a compromise between ashlar and rubble masonry. Most of the stones are rough-dressed (which still qualifies as rubble masonry), but the playful arch and the amazing stone that makes the watershedding transition from wall to chimney are ashlar stone. The mortar joints in this wall are red, and they are convex—rather frisky for such a conservative medium.

Tools for Quarrying

Ashlar stonework begins at a quarry. Normally, a desirable type of stone is quarried where it is most accessible. Blasting for building stone is only done to remove layers of unwanted dirt and stone. The shock is detrimental to the stone itself. For that matter, dressing the faces of a stone with a hammer can damage the rock by setting up internal cracks. The preferred method of obtaining building stone is by

Fig. 13-1. This playful stone building has a mixture of rubble and cut stone.

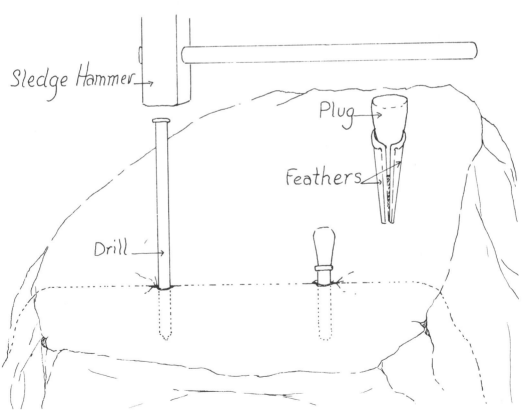

Fig. 13-2. Tools for quarrying hard stone.

splitting off big chunks at a natural bedding plane, using sledgehammers and wedges and steadily working the crack until the top comes loose.

With unstratified stone, like granite, a fissure must be made. Four tools are needed for this step (FIG. 13-2). The drill is nothing more than a hardened steel bar with a star point, and is held by one man while another smacks with a sledgehammer. With each blow, the drill man gives the tool a quarter turn. Gradually a deep hole is made. This work isn't much fun for the hammer person or the drill holder. The other two tools are used as one and are called the *plug and feathers*. The feathers, in two pieces, are the temporary metal sides of the hole. The plug is a tapered steel pin that is hammered into the feathers that line the hole made by sledgehammer and drill.

A whole row of holes is made where the quarrymen want the hard rock to break. Then each is filled with the feathers and plug. The quarrymen drive the heavy hammers on those plugs in unison until the chunk breaks. The process can be repeated with the newly released piece until it is in a more useful shape for further dressing by a mason.

Tools for Dressing Stone

The next step in obtaining ashlar stone, after getting the raw material loose from the bedrock, is to rough-shape the pieces with an axe or a pick. An axe is much like a two-headed wood ax, only much blunter (FIG 13-3). The roughly square piece is then improved with a hammer and pitching chisel. First, the corners are established, normally with a toothed chisel. Next, the margins of each face are drafted with a toothed chisel (FIG. 13-4). Each side of the stone is treated this way, unless the back is allowed to stay rough.

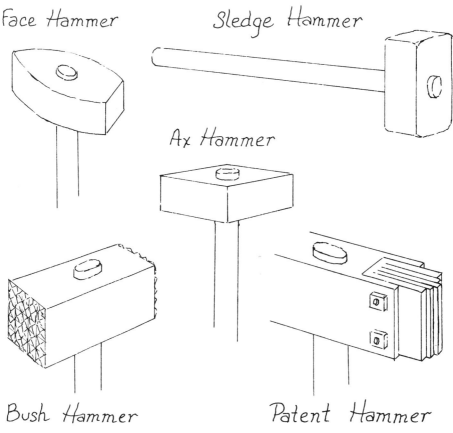

Face Hammer

Sledge Hammer

Ax Hammer

Bush Hammer

Patent Hammer

Fig. 13-3. Hammers used in stonemasonry.

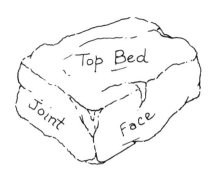 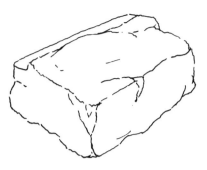

1. Rough squared stone

2. Pitch off the 1st edge

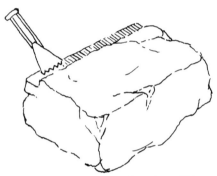 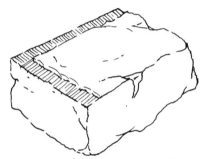

3. Cut the 1st draft

4. Pitch off and draft the 2nd edge

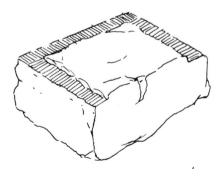 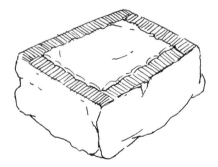

5. Repeat with the 3rd edge

6. Repeat with the 4th edge

Fig. 13-4. Drafting the margins of a rough quarried stone.

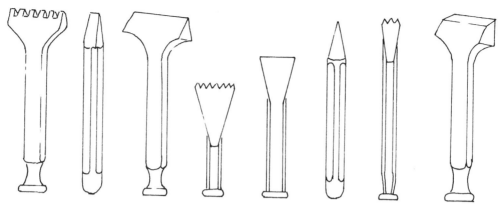

Fig. 13-5. Chisels used for shaping stone.

A stone has a face and a back, a top bed and a bottom bed, and two sides, which are called *joints*. Frequently, on the face of the stone, the rough aspect within the toothed margins is left alone. When it is removed, as it would be on the other sides of the stone, it is done with a point if the stone is very hard, or with a straight chisel if the stone is softer. These untoothed chisels are called *droves*, or tools. The names of the tools can vary among locales. FIGURE 13-5 shows various chisels for shaping stone.

Finishes finer than what chisels can offer are made with hammers, like the patent hammer or the bush hammer. The bush hammer is something of the meat tenderizer sort of tool. A peen hammer will give a stone a uniformly pocked surface. For even finer finishes on stone, abrasives are used. Sedimentary stones with distinct and parallel bedding planes can be split using lifts (natural cracks) to give stones with essentially finished top and bottom beds. This much (top and bottom bed) of the job of making uniform blocks of stone is often done by nature.

Usually, only the rough dressing and pick work is done at the quarrying site. For hammer and chisel work, the stone goes to a *banker mason*, who works at a stone cutting bench called a *banker*. Banker masons also carve geometric patterns in stone. When a nongeometric pattern, a figure, or writing is to be included on the stone, the job leaves the realm of masonry and goes to a carver, or artist.

NONSQUARE BUILDINGS OF STONE

Making curves with stonework is quite feasible, if the curves are gradual and the stones are fairly small. Bat-ter boards and strings are pretty useless on curved structures, but there are other ways to lay out the outline of a structure.

To make a round stone building, for instance, you can set up a plumb post with a ring over it, and tie a nonstretching rope, or wire or light chain to the ring. If you walk around the post holding the chain out tight and level, the end of the chain will make a circle. You can use the end of the measure to set the stone faces. The post can have notches or nails every 12 inches in altitude. The ring will tend to stay in the groove. If you need to brace your post to keep it plumb, you will need to walk the chain under the braces as you move around the circle.

You can mark the thickness of a round stone wall by the chain. The outermost point would be the outside face. Mark the inner face 18 inches in from that, or whatever thickness you are using. If the post stays plumb, the line doesn't stretch, and you hold the line fairly level, the end of the line will always fall on a point on the outer surface of the stone cylinder. If your post is tall enough, you can mark the top height of the wall on it, and a level radius from there will show where the capstones go.

Vaults and Domes

Masonry can be used to vault over spaces without internal supports, even temporary ones. The stone huts of Perigord, France (FIG. 13-6) are good examples, as are the trullo shelters of Cisternino, Italy (FIG. 13-7).

To give yourself faith in the possibilities, try the following experiment with dominoes. Arrange 6 domi-

Fig. 13-6. A primitive stone shelter with stone roof, of the type found in Perigord, France.

noes on edge lengthwise to form a hexagonal enclosure. Imagine that the hexagonal space represents a living area. Without interfering with that altitude of space, lay 6 more dominoes, this time lying flat, on top of the ones on edge. You need to let part of a domino stick out for them to fit. Next, crossing the joints below, lay another circle of flat dominoes, this time in a little closer and using only 5 dominoes. Then do another course with 4 1/2, then 4, coming in toward the center. Then it's 3, and 2, and finally, a single stone at the top. The top domino in this illustration, if it could fall straight down, would be completely within the walls of the miniature shelter. It has no internal supports, yet I was able to balance a quart of water on the top domino. The toy building supported a load on its roof that exceeds the weight of the building, yet there were no fasteners at all.

With the recent improvements in passive solar technology, vaults and domes could become realistic possibilities for residences once again. The great mass of concrete, brick, Gunite, ferrocement, or stones could be put to double duty as the mass in a thermal storage system.

The other big plus of the dome shape is its ability to support great weight. In the past, this ability hasn't been used to advantage. The domes and cones supported only their own weight. With the advances in insulation and waterproofing, masonry domes could now be insulated on the outside, waterproofed, and earth-bermed. As many openings as structurally possible would be on the sunny side of the structure.

As far as getting the most structure for the least materials, you can't beat a sphere, in theory. You get more volume of space enclosed for the least amount

Fig. 13-7. This is what the ceiling might look like inside a typical trullo house of Cisternino, Italy. The stone is frequently laid dry.

of surface area. If the surface area represents the materials, you not only get more house for the materials, but it is much stronger because of the shape.

A hemisphere of masonry is a charming possibility. You would get the same efficiency of materials, and a flat floor, too. A brick hemisphere compared to a brick box would be easier to build, a quantum leap stronger, able to shed wind and rain better, have less surface area to lose heat from for the same amount of space, and perhaps best of all, could be earth-bermed and be able to heat and cool itself. The bigger the hemisphere gets, the more space you enclose for the amount of bricks. The deal gets better with size.

The layout for a masonry unit hemisphere is charming in its simplicity. There is but one measurement you need: the radius. Suppose you wanted a 15-foot-high masonry hemisphere. A length of chain 15 feet long and staked in the ground would touch the face of every brick in the masonry hemisphere. This is so primitive, you really don't even need a measurement. You can just guess how high a ceiling you want at the top and use a length of rope that long.

You would need a footing before laying a hemisphere. It should be as any footing, wider than the wall it supports and down to firm ground below frost. You can scratch out the footing with a chain that is longer than the radius by the desired amount. For the

210

inside wall, you can scratch another line with a chain shorter than the inside wall by the same amount. Then dig a hole plumb down from those concentric circles.

The first course would lie flat, on the circle drawn by the pencil on the end of the chain as it rotated around the center hook in the ground. The next course would angle up slightly, to meet the end of the marker chain, with each brick face making a right angle with the chain. The last brick, in theory, would be straight up from the center point in the ground, with its face still 90 degrees from the chain line.

With extra-light bricks and sticky mortar, you can build such a house without support forms of any sort. The work would need to be done from the inside, on scaffold that didn't touch the walls.

You can make openings in the hemisphere, after it is complete, by removing units. To achieve conventional spaces inside a hemisphere, use internal wall divisions. Lofts can be supported from bolts in the masonry.

Spanish bovedas, or vaults of brick, have been around for centuries. Some of them are built without temporary support, as at architect O'Neil Ford's Texas boveda house. Here the vault was laid without portland-cement mortar or forms. (See *Fine Homebuilding* magazine #23, p. 28.)

Cones

Cones of corbelled-in courses of stone or brick are another possibility. You can build one uniformly by leaving a temporary center post, and running a line from a stationary point at the top center of the post. Draw a circle at the footing. To check the stones, pull the string to that circle. It should be able to go around the cone without hitting any of the units.

Hemispheres and cones can be built on top of cylinders of masonry to get more useful living space. Even a vertical round wall has tremendous strength against pressure from the outside, compared to a square wall. The Spanish bovedas span across rectangular spaces, and are much more complex to build.

An egg is a masonry container, built inside a chicken, mostly of lime. It would save the bird on materials if the egg was perfectly round, but it makes for a tougher job of laying. Have you ever tried to crush an egg in your hand? A Gunite-dome house can support a load of earth up to 27 feet thick, yet it is like an egg shell in scale.

AESTHETIC MASONRY

Objects of art can be added to stone walls for an even more interesting wall. It is easy to overdo it, however. My rules to myself about using foreign objects in a stone wall follow:

☐ The object must be fairly inorganic, hard, and able to handle the weather. Likely candidates for inclusion would be noncorrosive metals, glass, dry bones and teeth, shells, ceramics, and minerals of interest. Avoid wood, plastic, rubber, or anything that is going to become something else in the near future.

☐ The object must have enough depth to be held tightly by the mortar. Notice the bronze horse in FIG. 13-8. Its rear legs and tail are tied in the wall. When the wall is pointed, the horse will be hooked in even better. The wrong approach would be to merely press a coin flat on the surface of the mortar. It would fall off soon.

☐ The object should be used as a filler, not a stone. Structurally, the wall shouldn't depend on the art object. The curios should be used to break up a large mortar space between poorly fit stones. Stones shouldn't have to work around the art.

☐ The objects should not dominate the wall; rather they should be gradually discovered by the casual observer.

FIGURE 13-9 shows a close up of an arrangement of rubble stones under an arch and over a mantel. A bronze monkey fills the otherwise overly large mortar joint. Notice how the monkey is treated like a small fill stone with a mortar joint around it. I got the effect by first packing the space with mortar and then pushing the clean art object in the mortar. Using a small trowel, I then force the mortar tightly around the curio. Later, the pointing mortar is dressed, and in this case, brushed with the rest of the pointing job. Upon close inspection, the brush marks can be seen on the mortar. The ashlar stone above the monkey has clearly visible toothed chisel marks on its margin.

Inside walls are more forgiving on the type of object, although I would still lean toward heavy, earthy things that have a similar nature with stone. If you put

Fig. *13-8.* *Art objects in a wall should have enough depth to be grabbed well by the pointing mortar. This horse is held by his unseen rear legs and tail.*

Fig. *13-9.* *The bronze monkey is treated as a fill stone in a large mortar joint.*

a plastic toy in a wall, it will dominate the wall without really deserving to. FIGURE 13-10 shows an inside wall with many curios. The small turtle shell near the center of the picture is not in the wall very far and would not last long outside. At the bottom of the wall toward the center is a strange mineral nodule taken from a deep cave. It goes well into the mortar, but it is mostly iron and it would rust badly outside. If you have something in a wall that rusts, even ironstone, the rust will soak into the mortar and stain it. Water dripping off the ends of oak rafters will also stain a wall.

Notice how the white mortar in FIG. 13-10 looks almost like sand that the stones were pushed into. It takes extra time to point this tidily, but it doesn't

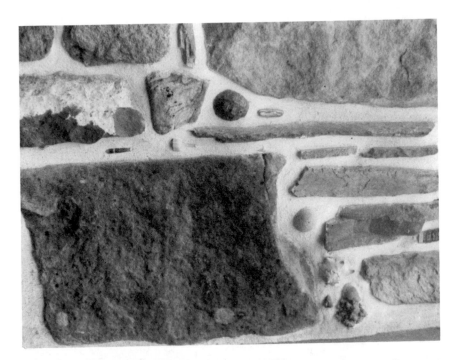

Fig. 13-10. Interior stone walls are more forgiving of marginally deep curios like the turtle shell in this wall.

Fig. 13-11. Light stones look best with dark mortar.

take muriatic acid. Interior walls come under much closer scrutiny than outside walls. Cement doesn't have to be ugly.

Color and Shade

FIGURE 13-11 shows a wall done with dark mortar. I think dark mortar looks best with light-toned stones. This section of wall has some white quartz and light-toned sandstone. They stand out nicely, but the dark ironstone at the bottom right doesn't go as well with the dark mortar. The stone in the upper left has the same moss growing on it as when I plucked it from the forest four years ago. Extra care must be taken to keep vegetation intact throughout the transport, handling, and cleaning of stones. Notice next to the furry stone is a small whitish triangular piece that breaks up the large triangular mortar joint there. These little stones are a concession to less than perfect fitting, but that is inevitable when working with rubble.

The stones in this wall are fit as closely as possible with rugged rubble that has a complete mortar joint around each stone. Tighter fits are possible if stones are allowed to touch each other. Oddly, this stone wall was laid in regular concrete. The black mortar is less than 1 inch deep.

These mountain sandstone suggest quite a lot more life, color, and motion than quarried stone, which has no individual identity. It's like the difference between a log and a 2 × 6. A fine ashlar wall cannot compare visually with the organic shapes of a well-fit rubble wall.

The suggestion of motion is more pronounced in the wall in FIG. 13-12. The bottom half of this playful, yet stable, wall appears to be swimming to the left. This wall is about as chaotic as a coursed rubble wall can get and still offer a feeling of stability and strength. The stones are of every possible kind, color, size, and shape. White mortar is best for showing off the color in a stone wall. More light is reflected off the mortar to show the hues of the stone. (See Color Section.)

Other Aesthetic Uses of Stonemasonry

Sometimes stones are used by themselves to suggest a certain feeling and to give a particular accent to a yard. FIGURE 13-13 shows part of a retaining wall in a steep yard. The wall has accommodated a tree within its boundaries, as well as two other stones. The single stones seem to be renegades from the wall. Minor artistic statements like this are the easiest of masonry jobs.

Fig. 13-12. Several fish-shaped stones in this frisky wall suggest motion toward the left.

Fig. 13-13. *Solitary stones can be used for artistic accent near stone walls.*

Oriental rock gardens take that artistry to a much deeper level and might make perfect sense in an area you want to observe and maintain as something of a moonscape. I have done a lot of this kind of work for people who don't want a lawn to mow. A rock garden is a viable alternative to growing grass. Whole books have been written on the subject of arranging nonfunctional stones. I can't accept that there are any rules about artistic gestures, however. I say experiment. Feel free to do what you like for the eye. Decoration shouldn't be toilsome or restrictive.

Cairns

A *cairn* is a stone pile or stack used as a monument or a marker. Often, when I gather stone for a job, there is extra when I'm done. I bring home the extra and toss them off the truck. Then, mostly to keep them off the lawn, I stack the stones into a quick pile. I've got several of them around the house and it amazes me how much people enjoy the sight.

FIGURE 13-14 shows such a pile. This people-sized pile contains about one ton of rocks and took about a half hour to assemble. This cairn hasn't blown over in three years.

I have older and larger cairns, which possess a lot of charm for something so fast and easy to build. Birds find the stone piles a convenient landing place and reptiles hide in the crevices. In the winter, we feed deer at one stone marker, which they can locate quickly from a distance.

In Scotland, cairns were used to mark boundaries and forks in the road. Mine serve no purpose except to keep stray rocks from making the lawn impossible to mow.

Working with Existing Stationary Stones

FIGURE 13-15 shows part of a retaining wall built to connect a series of boulders that were already in the yard. The wall accents the entrance to a home and saved the job of moving some huge rocks out of the way. The wall has a very organic feel, and seems to grow from the earth. Masonry here has been used to accent a natural obstruction.

In many cases, walls of this sort are as easy to build as it would be to get rid of the stones. Indeed, many of the early walls in the Eastern United States were built as a way to get rid of unwanted stones from

Fig. 13-14. A cairn of stone is fun to stack and interesting to behold.

agricultural ground. For this reason, walls are still sensible at a new construction site. What else can you do with clunky rocks lying about a new yard? Before you consider having your unwanted stones buried or hauled away, consider arranging them into a pleasant, organized structure in your yard or at the boundary.

LAYING BRICK
AS RUBBLE MASONRY

Bricks are delightful to work with if you are used to working with big, clunky stones. To me, a brick is an artificial stone of the correct size to be held by one hand. This is the big design breakthrough for production masonry. A brick mason can keep a trowel in one hand, and use the other to lay the units. This can't be done in rubble stonework, or even ashlar stonework. It takes two hands on the stone.

Over the last 10,000 years, bricks have varied remarkably little in size, except for one year in the 1800s in England, when a tax law was figured per brick on new construction. Almost immediately, bricks tripled in size. The bureaucracy of the time dropped that approach quickly, and bricks in England were back to the usual size the next year.

Bricks are made to be laid in very uniform rows with even mortar joints. This can be intimidating to the novice mason. I see no reason not to make thicker walls of cheaper bricks and lay them more like rubble stone. Many brickyards have reject heaps with

Fig. 13-15. Walls can be built to accommodate unmovable boulders.

216

bricks free for the hauling. The bricks in these heaps vary in size and shade. Many are broken. If these odd-shaped and -sized pieces are worked into a thicker wall with less demand on the perfect alignment, a beautiful, solid and cheap wall is suddenly within grasp of someone who would never consider laying bricks.

FIGURE 13-16 shows a rubble brick wall. This $30.00 wall took three days to build and put two tons of mass in the path of direct winter sunshine. There are vents in the bottom of this Trombe wall, and many nooks and ledges. There is no pattern at all, and the bricks are laid in all of their possible positions. Because these bricks were rejects, there is a wide range of colors and shades. You won't find uniform mortar joints or predictable alignment in this wall, but it is one of the most colorful masonry walls I've seen. It is interesting, but not threatening to the eye, and the wall serves a practical purpose (room divider, ceiling support, Trombe wall) at a bargain price.

I believe a good brick house could be made of salvage brick if the walls were about 1 foot thick. This would be much easier on the body than making a stone house with 24-inch walls. If you have any luck fitting rough stones, you'll find bricks laughably easy to stack into a wall. Think of the junk bricks as small stones, and you won't be tempted to use them in a thinner brick-type wall where the standards are too high for nonuniformity. If you aren't really confident you can keep things very straight and level, and if you are uncertain how the mortar should be put on the brick, you can make the wall thicker and punt. Chances are good you'll come up with something cheaper, stronger, and more interesting than a conventional brick house or wall.

I have built dry-laid brick walls that look and work great. I have an outdoor dry-laid brick planter that is intact after three winters. In the spring, I push the bricks back in line.

Much friskier things can be done with trash brick than the rather stodgy Trombe wall in FIG. 13-16. FIGURE 13-17 shows a random, curved brick planter which includes a set of steps. The plants in this one-day, $5.00 creation are low succulents that don't interfere with the sunlight entering this solar house. Curves, corbels, and bricks in every possible direction are a possibility, but don't forget about gravity.

MISSING PROJECTS

There are certain standard stone projects I have omitted from this book. Some are not here because of my own bias; some because of obvious trends.

Fig. 13-16. Reject bricks can be laid as rubble for some unusual patterns and shades.

Fig. 13-17. Rubble brick makes curves easy to build.

Barbecues

The big stone fireplace is on the way out. I suspect the backyard barbecue is soon to go. Not only are there new pollution standards emerging that will hurt the barbecue scene, there is mounting evidence that charcoal-burnt animal fat is some of the most dangerous food a person can eat.

Masonry barbecues, like fireplaces, are rather complex to build. Most people I know who have one also have a portable gas grill, which is what they end up using to cook out because it is mobile and very convenient. If this is likely, why bother with the difficult masonry? Barbecues are more difficult to build with rubble stone than brick. Also, stone is not very good as a material for a barbecue because of the heat changes imposed on the material. Brick is much superior for taking those temperature changes.

Stone Driveway Piers

Another project I dislike is those silly piers people build on either side of their driveways. Maybe it's only in Pennsylvania, but when I'm out on the road, I see dozens of these things along the highway that are heaved over or busted to pieces. I suspect people are tempted to skip the footings on something so

inconsequential, and they heave. The other problem is that people drive their cars into these masonry markers like crazy.

Stone Planters

Exterior planters of stone are common masonry structures that can frequently be found smashed to pieces. If you live where it freezes in the winter, an outside container for wet dirt made of masonry units will probably break from the freezing soil. It's a wonder to me how so many people are willing to completely ignore this phenomenon and build these things.

I took a short drive near my home to get a picture of a busted-up planter for this book, and I found nine of them to choose from. FIGURE 13-18 shows a round stone planter in a mortar mix of 3,000 psi strength that couldn't handle the strain. See how the wall chose to stretch where the joints were not lapped.

Another big problem with exterior planters is that they are persistently used in parking areas. If the frost heave doesn't get them, the cars will. Seven years ago, I was asked to build some stone planters in front of a roadside fruit market. The owner thought they would make the area in front of the stand more at-

Fig. 13-18. Exterior stone planters are often busted by freezing soil.

tractive to customers, as well as give him a place to put colorful flowers. I expressed regrets about building something that was fairly certain to fall apart. After a persuasive effort on his part, I agreed to do the planters, but only if they were filled with compost instead of regular dirt. Being an avid gardener, I realized that compost and humus was so light and full of spaces, that it could freeze without exerting much pressure on its container. When something fluffy freezes, it is easier for it to squash inward on itself than to push something heavy outward.

The planter in FIG. 13-19 has been intact for years only because it is filled with compost, and has drain holes and gravel in the bottom. While I was building this and another planter along the busy road, they were struck by cars three times. I kept fixing the damage, and finally the owner put railroad ties in the parking lot to keep cars from backing into them. The slightest tap of a bumper will crack apart something with walls this thin.

Another fundamental flaw in a masonry container for plants is that the walls need to be thick enough to accommodate the rubble, and that doesn't leave much space inside for plants. If the walls are a scant 1 foot thick and the planter is 4 feet wide, that only leaves 2 feet of space inside, with walls of questionable integrity.

Dry-laid planters make more sense, because they can move some and be pushed back in line occasion-

ally. Unfortunately, the walls would need to be even thicker to stand, which implies even less inside space, if you want to be able to reach across the planter to tend plants. In conclusion, exterior mortared stone planters have three strikes against them.

Interior planters are a different story. Hopefully, they are not in danger from frost or cars. Walls can be made very thin and can be treated more like a piece of furniture than a stone wall. More on this in the next chapter.

Fig. 13-19. This stone planter has resisted frost heaves because it is filled with fluffy compost, not heavy dirt.

STONE HOUSES

The paradox of the real stone house should start to become evident to the student of masonry or carpentry. The old-fashioned, 2-foot-thick walls make a house very hard to heat, unless it is well insulated. If you insulate a stone house, not only don't you get to see the stones on that side, but you must add a frame wall of sorts to support the insulation and its covering. In a sense, a conventional stone house has a wooden house inside. It is so much less work to just build the wooden house and skip the stone part.

If you build a free-standing wall with faces on both sides, would you want to cover up one whole side? If one of the sides of the wall is to be covered with insulation, it probably makes sense to build the walls against a temporary plywood backing. This method is explained in Chapter 7.

Insulating the outside of a stone wall is superior for energy efficiency, but that leaves the inside walls exposed. The trouble is that the outside insulation needs a weatherproof coating. With the insulation on the inside, no extra weatherproof layer is needed, although some form of interior wall is.

Another problem with insulating the outside of a stone house and exposing the inside stonework is that so much of the interior wall surfaces end up being covered with something. A stone face is too hard to make to have it behind a refrigerator. Who would want to spend several days assembling the stones that would only be seen at the back of the closet? The places where stones don't show can be done much faster in block or concrete, and it gives a much smoother surface for putting cabinets against. Look at the walls of the room you are in. What percentage of the surface of the walls is exposed? It is not worth the trouble of making interior walls of stone.

Stonework is not necessarily a desirable interior surface for walls. The dark surface is riddled with shadows and ledges to collect dust. Stone walls are rarely comfortable to lean a body against. Even at 72°F, a stone wall will chill you. A rough rubble wall is a bad place to bump your head, and a hard place to hang your hat. Rubble masonry rarely holds up to close scrutiny with good looks. It is one thing for a stone house to look great from the street, but when you sit at a desk in front of a rock wall, you tend to see more detail than that artwork should be showing. Next time you see a stone building or structure that you think looks good, go right up to it and look at the rocks and the mortar. I know it looks better at 30 feet.

A Sensible Stone House

With the advent of cement blocks and foam insulation panels, a stone house for the times might make more sense done the following way. Build a house of 8-inch block on an extra wide footing and have extra overhang on the roof. There are getting to be some wonderful blocks that are made to be dry-laid, and they have some insulation value. You can assemble these things like a child's toy. When they are in place, add a special surface mortar to the blocks that gives them some common bond.

When the house is up, which won't take long, gradually add an 8-inch stone facing on the outside of the block, but put a few inches of foam board against the block before you lay the stone. The thick facing will look authentic and add some structural integrity overall. The block wall inside will present a smooth, flat surface which can be finished with plaster or joint compound to look like a plasterboard wall. This makes a more livable interior wall that will reflect light well, and you won't feel badly about covering large areas of it with inevitable furnishings.

In this sandwich of masonry, the foam will be protected from the damaging effects of the sunlight and the weather. I think you could use the cheapest of foam panels in a situation like this. The foam will also be protected with a fire barrier, and so wouldn't constitute a health hazard.

The best part about a house like this, which with 4 inches of foam would have 20-inch-thick walls, is that you can get the block part up very quickly and start to live in the house while you finish the outside facing gradually, without pressure. True, you wouldn't have insulation until the facing was in place, but you could do quite well through three seasons of the year without any, by which time you could finish the facing.

Of course, a house built this way would be a lot easier to heat than a conventional stone house. In fact, with strategic openings for some solar gain, a house of this nature would compare favorably with any common housing. The mass of the block walls in this design would be within the thermal barrier of the house. This could work very much in the favor of the house as far as ease of heating, and be very adaptable to solar designs.

Fig. 13-20. This rubble stone building required more than a year of labor and is barely larger than a one-car garage.

I don't like the idea of splitting a rubble wall in half with a barrier of insulation. I know it has been done, but an 8-inch wall of rubble cannot be very stable. A pair of stone faces cannot be expected to do the job of a stone wall. A pair of 8-inch brick walls with insulation between would work well, and the walls would still be thinner than in an old stone house. If you can get free bricks, this would make a fine house.

Slip-Form Stone Houses

I don't think form stone houses make much sense. I don't think they look very good; I don't think they have much integrity or longevity; and I think they require too much work for a building that still needs interior insulation and wall finishing. Because you can't see the faces of the stones when you lay them in the forms, you can't keep track of what the wall will

look like when the forms are removed. I have yet to see a slip-form stone house that looks as good as a well-stuccoed block house, yet it takes a lot more work.

A double-sided form stone wall is nothing more than a concrete wall that is weakened by the inclusion of overly large aggregate: the stones. Individual aggregate shouldn't be anywhere near the thickness of a wall or slab. Slip-form walls break this rule. The stones make breaks in the concrete wall, yet they are not really stacked well enough to be independent of the cohesion factor.

If you are going to go to the trouble of setting up forms, why not set them closer together, and use fist-size stones and concrete to fill them. When the wall is done and the forms removed, you can parge the outside of the wall with an earth-tone mortar mix, giving a pleasant adobe look. There is an awful lot of

221

information about building stone walls with forms. There are dozens of techniques, many of them being patented, but I can't find the sense or the beauty in a slip-form stone house.

Don't let anyone tell you that stone is a good insulator. Two feet of stone isn't as good as 1 inch of foam. The problem of having a stone wall surrounding your interior walls in the winter is this: If it is below zero for a spell and your stone is icy cold, and then it warms to about 40°F, your house is still going to be surrounded by these heavy ice cubes. Your heating system, even though there is insulation, will have to fight against this load of cold mass. In summer months, the same phenomenon can keep your house too hot.

For outbuildings and uninsulated buildings, old-time rubble masonry might still make good sense. Both sides of the stonework would show, but who wants to put that kind of effort into an outbuilding when there is the option of fast, light cement block looming in the background?

FIGURE 13-20 shows a stone outbuilding that was built by two men of local rubble. The 18-inch-thick walls have show faces on both sides, and there is no insulation. Although the inside space of this building is not bigger than a one-car garage, it took two masons six months to lay the stones and build the roof. The job was stalled by winter, so it actually stretched out beyond six months.

This was one of the first stone jobs I worked on, and it made me realize what a stone house was about. For a year of effort, a one-room building that is impossible to heat was gained. It is charming and it might last 500 years, but it is hard to recommend as sensible in any way. This little building has over 40 tons of sand in the mortar between stones. If you have any fantasies of building something of this nature, it is good to be aware of the pitfalls and paradoxes, and to have some grasp on the quantity of work involved.

If it still makes sense for you to build a stone house, I suggest you keep it simple, with few openings and low walls. After 8 feet in height, stonework takes quite a jump in slowness. Don't judge how fast you will build a stone house on the way it goes on the bottom few feet. I'm not trying to dissuade anyone from building a real stone house or even a stone castle. Just don't start one thinking you'll be done in a year, or that you'll have a very comfortable place when you are done.

I think it is for all these reasons that true stone houses are no longer being built. If there is to be a resurgence of stone houses, it could be in earth-sheltered designs, where other building materials are hard to get and stone is plentiful. The next chapter has details of one such stone house.

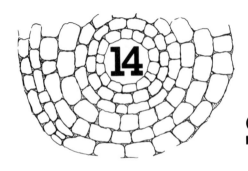

14

SOLAR HEAT & STONEMASONRY

PASSIVE solar architecture and stonemasonry are a natural combination, and one that I feel deserves extra consideration in this book. Among other reasons, people are using stonemasonry for its thermal mass. If you can get the sun to shine on it, stone will store some heat for later. Without the stone, the sun coming in will only heat the air uncomfortably, and almost none will be saved for later.

The idea with a passive solar house is to intercept some of the radiant energy that gets trapped behind glass, and gather it in a thermal mass of water, concrete, gravel, salt, or, in our case, stonemasonry. When waves of heat energy from the sun come through a pane of glass, they get trapped behind it and accumulate. This is why a greenhouse gets hot. If that accumulation of heat energy can't be stashed somewhere, the air will get too warm and the heat will be lost quickly when the sunshine stops.

It is also advantageous to concentrate the gain from the sun with reflectors. Reflective surfaces rob an area of much of the incoming heat and send it somewhere else. Heat goes from hot to cold areas. If it is warm inside your house and cold outside, the heat wants to go outside. If it is hot outside and cool in the house, the heat will want to get in. Insulation is used to impede that flow. It keeps the gathered Btus from slipping through the glass at night, to the cool dark air.

APPROACHES TO USING THERMAL MASS

Increasingly more houses in the United States are almost completely heated by the sun (including my own), and they are mainly of the passive type. The best of the bunch, as far as solar performance, are usually earth-sheltered and very well insulated with plenty of glass on the south side and some sort of movable insulation to cover the glass on winter nights. The other essential element of passive solar success is a mass to store heat. There have been many approaches to incorporating a large thermal mass in a house.

Btus

Btu stands for *British thermal unit*, which is the amount of energy needed to raise the temperature of one pint of water 1°F. That might not sound like a lot of energy, but consider the following. In Boston, Massachusetts, on the average day in January, over 500 Btus will strike a horizontal surface of 1 square foot. That's enough energy to raise the temperature of 500 pounds of water 1°F. Now suppose you have 100 square feet of masonry floor that gets direct sun on that average January day in Boston. There is enough energy striking that floor to raise the temperature of 500 pounds of water (about 62 gallons) 100°F. That energy would bring 62 gallons of 50°F tap water to 150°F, which would be enough for a long, hot shower. That average square foot in Boston figure would be triple in the summer.

In a sunny, southern city like El Paso, Texas, on the average day in June, 2,731 Btus fall on 1 square foot of level ground. Most of that heat bounces off the ground, and some of it soaks in the ground, depending on the nature of the surface. A flat, black surface can absorb nearly 90 percent of the heat it receives, while a white surface will only absorb about 25 percent. This phenomenon is fairly obvious to anyone who has ever walked barefoot on a hot tar road.

Passive vs. Active Gain

Anyone who has worked on a roof in the summer can tell you that dark shingles get too hot to hold. That direct gain of heat on the dark shingles is the easiest solar energy to obtain. If there are no gizmos involved, it is called *passive heat*. If you squirted water on your hot roof and the shingles heated the water, and the water went down a gutter spout into a heat storage, that would be active solar heat of the crudest sort.

One example of direct-gain passive solar heat occurs when the sun shines through a window of your house and strikes a dark, dense object, which absorbs the heat. An example of indirect gain would be if the dark, dense object was not in the path of the sunshine, but rather heat came to it through the heated air in the room. More typical is pumping heated air through a mass of gravel, where it gives off the heat it picked up from the sun. If there is a fan or pump in the system, it is called *active solar heat, with indirect gain*. My experience with stonemasonry and solar heat has mostly been in the realm of direct-gain, passive solar heat.

Hidden Aspects of Energy Efficiency

If part of the idea of an energy-efficient dwelling is to minimize the initial layout of energy, then massive, passive solar houses must be reviewed in that light, as well as by their performance after completion. I have seen supertight, superinsulated homes and envelope houses that could be heated with very little power, but the amount of energy used in their construction, when viewed with the life expectancy of the house, could make folly of the savings afterward. These factors are seldom seen as interrelated, but that doesn't mean they aren't, and any responsible approach to resource management will have to soon realize this.

Following is a simple analogy to help explain the point. Suppose you live in an isolated village, deep in a forest. The houses are all made of wood and they are heated by wood. One day someone discovers that you can make Styrofoam insulation and aluminum siding, put it over your wood house, and heat with much less wood. The rub is that you need to burn a tremendous amount of wood for energy to create these products. What if it took a 30-year supply of firewood to make the products, but they needed to be replaced in 25 years? The village, as a whole, would be better off not bothering with the new technology.

I doubt if any real figures yet exist for the amount of energy required to produce a given house in relation to the life expectancy of the building. The calculation also needs to include the environmental impact (the cost of repairing the damage to the environment caused by the production of the house) and the amount of energy required to live in the house. It is quite possible that some award-winning superinsulated houses require more energy to make than they save in their short lifetime.

All attempts at calculating such phenomena seem hopelessly biased. I know this much. A small, poorly insulated wooden house with an archaic heating system is more energy efficient than a sprawling mansion that gets 75 percent of its energy from the sun. The mansion will house two people, who are usually away, and the old farmhouse will have a family of six who actually live there. All attempts at figuring the impact of a structure on the environment must take into consideration how much service the structure provides. A second home, or a vacation home, from this point of view, is not energy efficient no matter what it is made of.

My bias is toward stonemasonry for houses whenever possible because of the longevity and minimal environmental damage. Following is a list of some common building materials and the amount of energy (in Btus) required to produce one pound of the material:

- [ ] cement: 3,755
- [ ] concrete: 413
- [ ] sand and gravel: 30
- [ ] aluminum: 112,676
- [ ] concrete block: 15,000 per block
- [ ] soil cement block (34 lbs): 170
- [ ] paper: 10,000
- [ ] glass: 11,000
- [ ] brick: 138
- [ ] rubble stone: 0

Another factor in evaluating the energy efficiency of a house is the people who live in it. Some people are willing and able to let the temperature of the house

224

drop to 55°F at night or while they are away. This behavior might well make a poorly insulated house more energy efficient than the latest solar place with all the gizmos that is kept at 72°F all the time.

I recommend a studied overview of the situation before you dig the footing for a new house. Keep in mind that there is some free heat available from the sun simply by orienting the long side of your house to the south. Sheltering the north and west sides from wind with a row of cedar trees is something most people can do. Also, at no additional cost to construction, you might add windows to the south side of the house and subtract them from the north side.

TROMBE WALLS

A *Trombe wall* is a massive wall, usually of masonry and 1 foot thick, that stands just inside a large expanse of sun-facing glass. Trombe walls of masonry are very effective in solar-heating schemes. There are ways to calculate how much mass you need. There are mountains of variables, foremost being the weather, but suffice it to say that too little mass and too little collector area is usually the problem. I don't know of a single solar house that is having a problem getting rid of excess heat at night.

After a certain thickness of masonry, probably 18 inches, there isn't much to be gained with more thickness. A Trombe wall is like a battery in a car. You only need a battery that is good enough to do the job and have a little reserve. You could get a 12-volt battery that would fill the entire trunk of a car, but it would be a waste of money and space because it would never really be needed.

FIGURE 14-1 shows a 24-inch-thick stone Trombe wall in construction. This wall is that thick because it also accommodates a fireplace and chimney on the opposite side. Building thicker and heavier won't hurt the solar heating. In the illustration, the wood frame in front of the big stone wall will have glass to gather

Fig. 14-1. An addition with a stone Trombe wall is being built.

heat from the direct sunshine, which will land on the stones. The glass wall frame is on a stone foundation that was laid in concrete. It is shown still unpointed.

The problem with Trombe walls is that they obscure the windows. You don't get to see out, and you don't get light streaming in. It all goes to the wall. One of the nicest things about other solar houses is the large amount of glass and the bright interior with a view. Trombe walls negate that.

Some passive solar houses and additions, like FIG. 14-1, have a greenhouse with a thermal wall between it and the rest of the house. This allows a bright space in the greenhouse, but still obscures the light coming into the house. Sometimes these walls include sections of glass to let in light. Sometimes the walls are water walls, either of barrels of water stacked into a wall, or with specially made tubes that go from floor to ceiling and stand side by side.

Water is better than masonry for holding heat. In fact, water, by volume, is more than twice as good as stone and three times better than sand for holding heat. The problem with using the massive wall of water for storing direct-gain sun heat is that it is often tricky to build, and difficult to work into the looks and feel of a living space. That certainly is the biggest gripe about walls of old oil drums filled with water. Frankly, I don't think people have had to work really hard on the problem yet, and if it ever comes down to staying warm, we'll see all manner of beautiful and functional water walls.

In my opinion, what makes a stone wall more reasonable than a water wall for stashing Btus is that it is beautiful to look at, it can be part of the structural support of the house, and it isn't a problem to make a rock wall thick enough to hold more than a water wall. An 18-inch stone wall will hold more heat than a 6-inch-thick water wall, with less design hassle. The wall in FIG. 14-1 is something of a compromise Trombe wall, with a 4-foot-wide space in front of the wall where a person can loiter in the sun.

SOME PASSIVE SOLAR HOUSES WITH STONE

Following is a brief review of some passive solar houses which I have worked on. All of the houses are relatively energy efficient. All of them have a hand-built quality and have been designed mostly by the owners. They all contain some scrounged and recycled building materials, and cost less than their more conventional counterparts per square foot of floor space. Some criticism is necessary to explain the advantages and disadvantages of the various designs.

Pickering House

The Pickering house is a true stone house, with laid, 18-inch-thick rubble walls throughout. Most of the walls were built off a plywood brace that was set plumb along a line on the footing and braced to stay there. The walls were laid while measuring 18 inches from the form to establish the face plane. This is a fast way to lay stones if one of the faces won't be seen. It also is a good way to get a very uniform back side of a wall that is to have insulation or waterproofing. This method allows an authentic arrangement and appearance of the stones that show, and a much improved structural integrity compared to slip-form stone walls, which essentially have the braced plywood on both sides. With the half-form, lines can be made on the plywood to show where the openings are, and to define the top of the walls.

The mortar used on this job was made of limestone tailings, or screenings from the production of gravel. It was mixed six parts to one of portland cement. This makes a very strong and cheap mortar. It is clunky to the trowel because of its larger grit, and it is pretty ugly to look at, but we left it recessed in the walls and pointed later with a nice sand and whitemasonry cement mortar.

FIGURE 14-2 shows the partially pointed interior walls of the house. The partition wall in the center of the picture has two faces that show and was built without the half-form.

The stones for this job (over 200 tons) were gathered from a rough farm fence where an apple orchard meets a forest, a mere 1/4 mile from the job site. The farmer wanted to expand his orchard into the woods, so the stone wall was in the way, and he was glad to have us take it away for free. Even free stone this close to the job site cost the owner of the project about $10.00 per ton in labor and truck harm. Perhaps a dump truck and loader would have been worth hiring, but we did it by hand with a beat-up pickup truck, which is considerably more beat up now.

The limestone tailings were brought in in 10-ton loads at half the cost of real sand. The mortar made from it allowed the cement to go twice as far because of the variety of sizes of grit, including large ones, and maybe because of the fine lime dust also present. We went through more than 80 tons of mortar, so the savings were significant.

The Pickering house was built on a south slope with a dramatic view of 30 miles. Trees shelter all sides of the lot, except the south, which has some deciduous trees for summer shading. The site location is ideal for solar performance. The house itself is built in three tiers that follow the hill. Each floor is staggered to the side of the previous one, (FIG. 14-3). The bottom level, furthest to the east, sports a solar greenhouse, complete with overhead glass. The second and third levels have vertical glass for the entire south aspect. There are three entrances: one through the greenhouse, and two on the east side of the house.

This house qualifies as an underground house because it is buried on all sides (except south) and the roof. Several retaining walls were required to hold back the dirt from the two entrances on the east side. They were not completed when FIG. 14-3 was taken.

Massive rafters that were needed to support the weight of the earth on the roof. One of the problems of the design, though an exciting challenge, was moving these 24-foot-long oak 6 × 12s. (See FIG. 14-4.)

Unique houses are not usually easier to build than conventional ones, and the Pickering house was no exception. It took some searching for a sawmill that would handle the big wood. The roof is decked with 2-inch-thick boards, covered with 4 inches of foam insulation, which in turn is covered with a rubberlike roofing sheet. The roofing had to be shipped in from a considerable distance and took some effort to locate. In a more urban setting, I suppose this wouldn't have been a design hassle.

The outside walls, after the half-forms were removed, were very uniform like concrete walls. As we laid the stones, we made an effort to keep the area behind the faces packed tightly without spaces. Pickering added treated wood nailers to the half-forms on the inside. To these he added many nails with the heads protruding in all directions, which would grab hold of the mortar as it was added. When he pulled the form after the mortar had set a few days, the 2- × -2 nailers stayed in the stone wall, held there by the nail

heads in mortar, and the forms came off, leaving the ends of nails (which had held the nailers on the form) sticking out of the flush-set nailers.

The purpose of the nailers was to have a way to attach foam insulation to the outside of the stone walls. The owner added 4 inches of foam to the outside of the soon-to-be-buried walls, using extra long nails, with extra broad heads. He then covered the foam with the same rubberlike sheet that covers the roof.

Before the dirt was filled in around and over this house, he laid a drainpipe at the base perimeter of each section. Evidence of the drainpipe can be seen in FIG. 14-3, where it is temporarily lapped on the roof. The exterior drain line was kept lower than the altitude of the interior floor on all levels, and each section of pipe was connected to the next lower section. At the bottom, a drainpipe takes the total water that has seeped down the walls at any place on the house.

To minimize the demand put on that drain and to lessen the amount of wet dirt adjacent to the walls, the owner/builder devised a gutter system to drain each section of roof before the rain gets a chance to seep down the walls. This gutter was made by curling up the bottom edge of the rubber roofing.

The real problem area of this house, in terms of water shedding, is where the different tiers of the house join. A condition exists wherein the roof sheds water toward the windows. In these places, the designer lapped rubber up the south wall of the next higher section. This rubber is shown in FIG. 14-4 at the solar openings of this upper level. The rubber sheds water, but leaves and dirt tend to accumulate against that overlapping section of wall.

The Pickering house performs very well for solar heating. The residents use no night curtains, and claim to use less than three cords of wood per year for all their backup heating needs. I would guess that with a movable insulation system, this house would need half that amount of fuel.

Some of the problems of this house are the same as its advantages. The place was built by hand, using ton upon ton of rough stone. It is a fairly big house (about 2,000 square feet of floor space) and it took a long time to build. These days, speed of construction is usually important because someone is paying rent on another place waiting for the dream house to appear. The Pickerings were fortunate to have a trailer

Fig. 14-2. The rough stone walls of this underground home were pointed after they were laid.

Fig. 14-3. The Pickering house before it is buried.

Fig. 14-4. An underground house is brighter than a conventional house if there is south-facing glass.

to live in while they took the time (four years) to build a stone house. When the house was done, they sold the trailer for what it had cost and it was hauled away. There is quite an advantage in living next to the building site of your new house.

In addition to the time and body-breaking work, this house presented a rather complex floor plan. There is nothing uniform about the size and angles of the zigzag of the solar house. The mastermind of the operation simply kept the forms coming, with the little nailers and the lines to show where the walls end. He had nothing more than napkin drawings to go by, but stone is forgiving of that type of creativity because it proceeds so slowly. You have a long time to think the thing through while assembling it.

Another problem with the design, which could also be seen as what makes the place special, is that the amount of wall surface area is very high for the amount of volume enclosed. The spread-out design has more surface than a boxier house that would enclose the same space. This makes a house harder to build, but more interesting to be in.

The dirt on the roof of this house is not quite deep enough to support a lush cover, or rather, it drains too well to keep the thin dirt above wet enough to stay green. This is not a big obstacle, and might be merely a matter of finding a more drought-hardy roof crop.

Another criticism, which is not really to the building but to the circumstances, is that several different masons got their hands in the job, with many different styles and degrees of skill. Because the walls on a big project are brought up in horizontal courses, the vertical progress is very slow. The bottom courses of a wall might be done by a different mason with different stones than the top courses on the same visual section. When you look at a big stone house, the part of the wall that was built within a narrow time frame is a horizontal band probably only a few feet tall, the whole way around the house. In the living room of the Pickering house, there are walls that had the bottoms done in one year and top courses done several years later.

It is hard to be sure over a long time span if the same people will be working or if you will have the same kind of stones. The rocks in this house are some kind of rhyolite that does not form or break on any regular parallel planes. The stuff was harder than granite to shape or break. The stones were too con-

venient not to use, but they made for some tough fitting and pretty wide mortar joints.

On the side of praise, this house will probably outlast any residence built in Pennsylvania in the last 10 years. An underground stone house has the potential to outlast even a conventionally laid stone house because the walls are lower and buffered by the dirt. It is hard to recommend a project this labor-intensive, and I'm not sure the owner-builder would have started the place if he had known how much he was getting into. You've got to want it bad, or fool yourself into doing something this excellent and outrageous.

If the Pickering house does last 1,000 years—and there is no reason it couldn't—that would make it one of the big-time winners in the energy efficiency race. In a way, the place is built of trash: stones that were lying around in the way. There is something marvelous about it, and a real stone house makes more sense with an earth berm and solar glass. What used to be the worst thing about a stone house in the North was the difficulty in heating. This house has turned that around by being dug into a hill and trapping sunlight on its rocks.

The Pickering house has required several outside retaining walls to allow the earth berm. These dry-laid stone walls let the structure nearly disappear into its surroundings. Indeed, this house is invisible from the approach, which is on the north side.

One of the biggest advantages of this design concept is that the thermal mass is also the structure of the building. There are no Trombe walls in this house. Even though it is almost completely underground, it has a very open, bright feeling because of all that glass on the south. Some parts of the floors of the house are 10 feet below grade, and when you tuck a house down this deep, there is no fear of the inside freezing or even going below 50°F. That temperature becomes the starting point, and any sun you can trap will bring it up from there. In most of the United States, a nonburied stone house is fighting against an exterior winter temperature of below freezing.

One of the unfortunate aspects of the timing of the closing in of the Pickering house is that the house was not buttoned up with windows and insulation until winter. In other words, the 200 + tons of masonry was ice cold when it was finally brought into the thermal envelope of the house. This was unavoidable, but if the place had somehow been closed in during July, those very warm stones would have let the house coast through the winter on the extra warmth.

Mudd House

Like the previous house, the Mudd House also was dug into a south-facing hill, with woods on all sides and some clearing on the south side. The Mudd house also has a solar greenhouse on the lowest (altitude) level of the south side of the house.

In the foreground of FIG. 14-5, you can see the block foundation for the wood and glass greenhouse to come. The stone wall divides the greenhouse from the house and gets full sunshine, which is stored in the mass. The 30-ton beauty has vents in it that appear at about mid-level in the greenhouse, but at floor height in the stepped-up house. Excess heat from the greenhouse is vented to the upper level of the house through ceiling vents in the greenhouse. A self-propelled circulation of air is the result.

The modified Trombe wall is outfitted with glass panes to let in the light from the greenhouse, while allowing the option of closing that area off from the rest of the house. FIGURE 14-6 shows some of that stone wall, which is visible in the living room of the Mudd house. The rectangular vent holes are evident at floor level. The wood frame of scrounged chestnut is held into the wall with bolts in the mortar. It is the frame for a pair of glass doors.

The north wall of this house, which can be seen in FIG. 14-5, is made of block. It has no openings and is buried on the outside, with insulation behind the block. This block wall was finished by plasterers to a smooth-as-silk finish. You would never know that it was a block wall in the house and not a regular plasterboard wall unless you tried to kick a hole in it. The block is 12 inches thick, with several of the cores filled with concrete. The extra strength and the extra mass are good in an earth-sheltered house. The white interior wall makes a much brighter inside than a stone wall would have.

Other stonework on this house include retaining walls that allow the earth berm to stay in place against the east and west sides (FIG. 14-7). The second floor of the house has no masonry, but it does have plenty of south-facing glass. The designer, co-owner and builder, Todd Mudd, used salvaged items in construction, including the building stones, which were pulled from an old foundation. Chestnut beams grace the visible framework, and poplar flooring gives local flavor to the house. The scrounging for this house was first rate and paid off handsomely. The house is worth quite a lot more than it cost, and it is a beautiful and comfortable place to live.

Like the Pickering house, the Mudd house requires about three cords of firewood per year as its only heat other than solar. The house also has no movable insulation system for the glass at night and would no doubt get by on much less fuel if it did. Both of these solar houses are surrounded by woods, so it is hard to raise enthusiasm for the hassle of window shutters when it is so easy to get another log for the fire.

Both houses have similar greenhouse design. I often wonder if the greenhouses are worth the hassle of building them. My criticism of the solar greenhouse is that the air heats too quickly and is tricky to regulate. The sloped glass is hard to waterproof and insulate. In the summer, it lets in sun that needs to be kept out. I wonder if these spaces make as nice a place to linger as was once expected.

The main advantage of the sloped glass in these solar greenhouses is for plant growth. Vertical glass will work fine for heat gathering and is easier to build in and insulate. Vertical glass won't allow summer sun to enter the structure if there is a small overhang. If you aren't doing major-league gardening, these spaces might not warrant the hassles. They are fairly unusable in summer, and not needed then for plant propagation. If you are only a minor-league gardener, you can start a lot of plants in a solar house with vertical glass only, and the heat gain will be easier to manage. Another problem is that, when you have glass overhead, nighttime insulation becomes even more important for keeping the solar heat in the building, but it is harder to insulate the overhead glass. Heat will pass through a skylight 20 times faster than through a stud wall with 3 1/2 inches of fiberglass insulation. What is the point of catching heat from the sun if it is going to escape at night?

A minor problem that occurred on the Mudd house, and something that often happens on stone jobs, is that part of the rock wall got covered by the concrete wall in the greenhouse. This certainly presents no structural problem, but merely implies that the contractor paid for stonework where concrete or block would have sufficed. Stonework is so slow,

Fig. 14-5. *This stone wall is a heat storage and room divider for a solar house being built.*

Fig. 14-6. *This is the back side of the wall in Fig. 14-5. Notice the vent at floor level.*

Fig. 14-7. *The block wall of this solar house is covered with foam, and then parged with ferrocement for an adobe look. The stone retaining wall keeps the dirt against the rest of the west wall.*

you might as well get a clear bead on what will eventually be exposed and what won't, and don't bother with fitting stones where they can't be seen.

FIGURE 14-7 shows the west end of the Mudd house. Todd's ferrocement work enabled the exterior foam insulation to be covered with a layer of attractive, weather tight masonry in places where the backfill dirt didn't cover the whole wall. The metal mesh and mortar is incredibly strong if done well. The bathroom of the house is graced with a ferrocement bath and shower stall that is as smooth as plastic.

An alternative approach on this west end might have been to build the retaining wall higher, and backfill up to the top of the first floor. This method would provide better insulation than the wire mesh and mortar over the foam, without much more work, if any.

The stone retaining wall shown in FIG. 14-7 was the builder's first stonework. I think he thought it was going to be harder than it was to build and was reluctant to go any higher. At any rate, he did an excellent job on both the dry wall and the ferrocement pargeover foam that insulates the block walls of the first floor.

The retaining walls of the Mudd house are an excellent example of practical stonemasonry. The walls hold the earth against the side of the house, which helps to keep it warmer. They are lovely to behold and were built by a novice stone mason in a few days with no money input. The wall has a gentle curve, done by eye to fit the circumstances. The stone walls—there are two outside this earth-sheltered house—act as protective arms, sheltering a patio area in front of the glass greenhouse.

Nunamaker House

The Nunamaker House is a passive solar house also on a south-facing hill with woods surrounding the site and some clearing to the south. Just having this kind of site puts you a long way ahead in the struggle for warmth. Nunamaker found one of those places on purpose, as did Mudd and Pickering.

FIGURE 14-8 shows the large expanse of glass on the sunny side of the house. Most passive solar houses are spread out lengthwise on an east-west line to get the most collector area for the floor area. If the house were deeper, the glass on the south would have to collect heat for more space. Notice the heavy drapes on some of the glass. They hold in heat at night and on overcast winter days. On partly cloudy days, it is hard to tell what to do with movable insulation or drapes. The light you get with the drapes open is worth a lot, even if you are losing some heat. I took this photograph on a partly cloudy day.

In the lower left of the FIG. 14-8, part of a stone retaining wall can be seen. This home is graced with many tons of retaining walls, as are many solar homes. Many solar homes are on steep south-facing

Fig. 14-8. The Nunamaker house has ample south wall glass.

hills. Living on a hill in itself is a good reason to have stone retaining walls—to create level areas. Many solar houses are earth sheltered, and this one is too, partially on the north side. The earth berms usually need a wall to keep them in place, or to keep the dirt out of a doorway. This place has retaining walls of stone for both reasons.

This house also has a rock garden, three sets of stone steps, an interior stone wall to back a woodstove, and a stone chimney. The chimney is visible at the far left of FIG. 14-8. It is made of the same mountain stone as the retaining walls. Stonemasonry continually pops up at solar sites, both inside and out.

The north wall of this house and the insulated slab, which is covered with terra cotta, constitute the main thermal storage in this house. Using a shed roof design like this and placing the glass on the high side, the light will penetrate deeply enough to strike the rear wall of the house. Naturally, there will be a lot of obstructions in the path of the sun. There seems to be a tendency among solar fanciers to keep their passive homes fairly uncluttered, so the sun can strike the mass. I consider this to be one of the side benefits of solar living; you don't need to drown in a pile of products.

Another hidden benefit of solar houses is that they are conducive to plant life: a moderate temperature, a place that won't freeze when you go away, a house without the dehydrating effects of a hot-air heater, and mostly a space that is very bright compared to the usual. Almost all the solar houses I've seen are big on plants and small on furniture, and I love it.

The highest glass on the house can be tipped forward to vent excess heat. The overhang of the roof on the south side prevents summer sun from entering. This is another neat thing about most passive solar homes: they stay fairly cool in summer. It is not just a heat trick, although my part of the country has very little of the too-hot problem. This house also has a wood stove backup, and reportedly uses about three cords per winter, since much of the glass is not outfitted with night insulation.

Typically, north and west walls have no openings in a solar house, because there is no direct gain to be had from those directions (in the Northern Hemisphere). Because the driveway enters from the west, this house has some small windows (on either side

of the chimney) on that side. It is nice to be able to see who is coming without going out a door.

A criticism of this fine house is that it is short on thermal mass, which means the air inside will get too warm and need to be vented. If heat is being vented, it isn't stored for later. It is tricky to have the structure of a solar building be the mass in a thermal-storage scheme if the walls are mainly wood. It is hard to keep a masonry floor exposed enough to the sun. A mere throw run on the wrong spot on the floor can cost you thousands of Btus per day. The backup heat system makes up for these solar compromises to comfortable living.

Perhaps the toughest part of working a solar heat plan is that somebody needs to be living in the house. So many of us are away so often it becomes nearly impossible to operate the controls of a house for optimum sun heat.

Bailey House

The Bailey house is a more or less conventional frame house with some added passive solar features. Again, the broadest part of the house faces south, and that's where most of the windows are. The house has a fully insulated concrete slab for a subfloor, with a stone floor on top. The outside of the footings is insulated and has a drainpipe. The insulation outside the footing ensures that the heat held in the slab will not slip away through any uninsulated masonry. One area of the floor, directly inside a large expanse of south glass, has stone imbedded in the concrete to create a rough flagstone effect. The owner-designer of the house asked me to build a simple stone bench in the sunniest spot of the house, with hopes of gaining some thermal storage and a crafty place to sit and set plants, etc.

FIGURE 14-9 shows the beginning of the solar bench in the Bailey house. The 18-inch-high wall is an outgrowth of the stone floor, done with the same type of stone.

The layout of the solar seat was a piece of cake. On an inside job, there are usually good places to attach strings. All the measuring can be done off a level floor, or down from a level ceiling. In this case, with unsided frame walls beyond the bench (FIG. 14-9, TOP), I could attach strings to a horizontal board, which I nailed between studs at the correct height from the floor, and level.

The little wall was to be 18 inches thick, so I made one mark on the bottom of the wood wall along a string that follows the front edge of the wall. Using a level, I made a mark plumb from that one on the floor to the horizontal board. Then I measured over 18 inches and got the other point. I put nails in the level board on the side unseen, hooked strings on the nails, and pulled the strings over the board in a small saw notch so they would stay put. I made the same type of batter board at the opposite end, and strung the lines.

The two strings represent the top edges of the bench. I made a pair of plumb marks from the strings to the slab and struck chalk lines along the slab, directly beneath the strings. That gave me a pair of parallel lines to sight across to sight the plane of the wall faces. For the top of the wall, I could sight across both strings to see if anything was sticking up too high. I set the ends of the wall by level and eye. I didn't want to run strings in the other direction just to set a very few end stones.

In FIG. 14-9, there are two face stones that stand above the rest. Between them are some fill stones. I mortared over the fill, and made a mud bed for more fill stones to bring the middle zone to the same height as the faces. A two-faced wall with rough stones is hard to fit in 18 inches. This bench also has the top and end stones showing, so the wall is almost completely face stones, with many stones having faces on two planes, and the corner capstones having faces in three planes.

The stones on this job were already picked for me and it made for quite a challenge. The 14-foot-long wall took three days to build and weighs nearly 3 tons.

In FIG. 14-10, the end and the top of the short wall is near completion. You can see that the string makes a straight line and the stones follow, with some parts in and some out of the line. In more traditional rubble masonry, no stone would have any part beyond the line. I like to average the dips and the bumps in a wall.

If you do let part of the rock go beyond the line, make certain it isn't touching the line itself. This will start to push the string into error. In FIG. 14-10, notice the small pieces that have filled the spaces created between larger face stones. It is best if the small pieces are not exposed on the side or face of the wall. A small stone is much more vulnerable on

top edge than on the top middle, surrounded by other stones. Even though a stone is small, it deserves the same size mortar joint as the larger stones. I point this out because there is a tendency to scale down the mortar joint with the scaled down stone. If you jam a little stone in a space and it has no slack, you won't be able to point it right. The stones shouldn't touch each other.

I laid the stone bench in type N mortar—one part masonry cement (1/2 portland and 1/2 lime) and three parts sand, with enough water mixed in to make a workable goo. I kept the mortar recessed as I laid the wall to leave room for a black pointing mortar.

In a wall this low, you need to plan for the capstones almost at the start. I set aside the smoothest of this rough bunch before beginning. I built up the place where they would lay by measuring down from the string the distance of the thickness of the capstone plus a mortar joint. If you try to fudge it at the top, you will have three problems. You will have stones too thin that come to the correct height; you will have thick stones that go too high; and you will have stones that aren't high enough. I saved the best stones for the spots I thought might actually be sat on.

FIGURE 14-11 shows the solar bench with the black pointing mortar roughed-in. For soaking-up sun heat, the black mortar has an advantage over white.

Black mortar is more difficult to use without staining the stones. If you use the stuff, you need to make an extra effort to be tidy. You could easily end up with a smeary looking mess if you aren't careful. White mortar is the least staining and most forgiving color for stone or brick work. When you touch black cement, your fingers will be blackened. The pigment takes some elbow grease to remove.

One of the nice things about the little sun niche in the Bailey house is that the wall is meant to be exposed to the sun, and won't likely get covered over with all manner of objects that would nullify the purpose of not only the thermal mass, but the looks of the stone.

In the same large, open room with cathedral ceiling and the stone bench, there is the stone chimney and hearth seen in Chapter 8, FIGS. 8-12 through 8-14. Unfortunately, that structure, weighing almost six tons, is not in the direct path of the sun in winter. Still, the contrasting stone structures on opposite sides of

Fig. 14-9. (top left) Interior walls can be laid out with lines from other walls.

Fig. 14-10. (top right) The top layer of stones is planned to stop at the strings.

Fig. 14-11. (bottom left) The black mortar is roughed in on the solar bench.

the room provide a very pleasing picture, like a matched set of different-shaped furniture.

Paulson Addition

This solar project in stone is more extensive and complex than the others. An old (150 years) frame farmhouse was to have a solar-heated addition, incorporating a greenhouse. The framework for the greenhouse sits on a stone foundation. There are many angles in the work, as well as three sets of stone steps and much interior stonework. Inside the narrow greenhouse, a mere 4 feet beyond the glass wall, is a Trombe wall of stone. The stone wall is 2 feet thick, 16 feet long, and counting the unseen stone foundation, 12 feet tall. This project contains over 60 tons of stonemasonry.

This Trombe wall is shown in progress in FIG. 14-1. The wooden wall with the windows in the background of that picture is the original exterior wall of the old farmhouse. When the addition was completed, this wall was removed, and the house was opened up to the addition.

FIGURE 14-12 shows the Trombe wall finished, with its big squarish limestone and black pointing mortar. This wall was bedded in concrete. To add considerably to the complexity of this job, the opposite side of the Trombe wall contains a fireplace and chimney, which extends another 12 feet higher than the top of the Trombe wall. From foundation bottom to chimney top is over 24 feet on this job. The wall is blessed with many art objects and curios, as well as vent holes and light tubes with crystals at the openings.

The owner of the project is a sculptor, and he designed the stonemasonry as an artist might. Practicality was not at the forefront of this project. This is simultaneously what is good and bad about this job. In stonemasonry, you pay dearly for the altitude, the angles, and the openings, especially when both sides of a wall are visible.

FIGURE 14-13 shows the project from a distance. The chimney is an extension of the Trombe wall. There is a thermal problem in this arrangement because there is no way to insulate a masonry chimney to keep its stored heat within the thermal barriers of the house. In other words, the stone chimney will leak heat by radiation and convection that is stored in the Trombe wall.

The limestone for this project came from an old barn foundation. Many of the stones were rough-dressed into a fairly square form. I thank those people of the past for their part in making this degree of integrity possible today. I certainly wouldn't have been paid to shape the stones like that, and it is nice to work with them.

This addition has some major bug-a-boos concerning energy efficiency. The stone foundation for the glass wall is not insulated, and therefore is another area of considerable heat loss. More significant is that there is too much mass for the amount of sunshine that can strike the wall with the roof sloping toward the south, rather than the other way, as in the Pickering and Nunamaker houses. As the sun takes its low winter arc, it only strikes the lower half of the Trombe wall. The addition has skylights to add to the heat loss. The skylights are necessary because the Trombe wall blocks most of the light from the greenhouse.

This structure is very pleasing to be in—very bright and creative, with cherry boards inside and cedar outside. My criticism of this addition is only of the solar aspects. Why let solar energy dictate any of the design criteria if you aren't really going to come out ahead in terms of Btus? I suspect that there are many supposed solar structures that lose more heat than they gain.

The problem areas in heat loss are hard to remedy without sacrificing aesthetics. Skylights are a great example. It is nice having the extra light in the daytime, but at night nothing is gained from a skylight, and plenty is lost. Winter nights are considerably longer than the days, so the skylights are losing the most when it is cold out. In summer, skylights bring in unwanted heat from the overhead sun. Summer sun is three times as potent for heating as winter sun, so it doesn't take much glass in the path of the sun to add many unwanted Btus to your interior.

Windows and doors on the cold and windy side of the house are another problem when you are trying to get free heat from the sun. It is hard to say just how far to go with your building plans to accommodate the sun. I think a lot of the free heat can be had without devastating the looks of a place. It is darn hard, east of the Mississippi and north of the Mason-Dixon line, to get all of your heat from the sun without making severe design concessions.

Fig. *14-12*. *This stone Trombe wall obviously has much more surface area than a comparably sized block or concrete wall.*

Fig. *14-13*. *The Paulson addition contains 60 tons of stonemasonry.*

Kellet House

The Kellets are a delightful family who took on the group project of building their own home. They built a passive solar, earth-sheltered house, mainly from salvaged barns and fine old buildings. The north side of the house is concrete. It is insulated and water-proofed on the outside, and earth bermed in the finest solar tradition (if there is one). The other walls are wood framed with barn timbers and siding. On the south side is a greenhouse with brick floor for thermal mass.

An interesting variation the Kellet house has is that the glass greenhouse panes are made to be removed in the summer and put back in the winter. Although the double panes of 6-foot-high glass are pretty heavy and hard to move, it only has to be done twice a year and only takes an hour, so it is worthwhile.

The family built with no construction experience. The job site was always humming with family and friends, with goats and dogs. The job they did was incredible, and the house unique. For one thing, because there was so much scrounged stuff from yesteryear, the place came out looking like an antique the minute it was done.

None of these old products will fit into today's 4- × -8 schemes, so if you use them, you are bound to get a very different feel, shape, and size. Materials impose their will on a job. With cement blocks, you get right angles and straight walls. With plywood, you get 8-foot walls on 16-inch centers. With old barn boards and beams, with stones from centuries past, with wrought-iron strap hinges and doors that weigh 400 pounds, you get a house that is not run of the mill.

The family enjoyed the task so much, they found themselves wanting to do it again. The only way they could was to sell the first house, which they did. The second house, which I did solar stonework for, is much like the first: passive solar with a greenhouse on the south. Again, the panes are set in for easy removal in summer. The house is built almost entirely of old building parts that were scrounged from all over the county—at sales and auctions, at demolition sites and abandoned structures.

I located the stones that were added to the new house plan at a very old and dilapidated outbuilding. The limestone chimney showed no mortar mess at all, and was standing beautifully next to its crumpled building. Rock hound that I must be, I stopped and asked the owner if he would part with the stones. He had intended to push them in the hole made by the building's basement so I could have them as long as I could get the hole filled with something. It took some telephone calls, but I finally found a way to get the hole filled and those stones to the job site.

The second Kellet house is something of a two-story solar saltbox. There are no north openings and few on the east and west. The construction is of timber frame on a massive slab of concrete. The 20 tons of stonemasonry I added, with the help of Paul Kellet, is in the form of stone piers between glass panes at the house wall inside of the greenhouse. The greenhouse wall is all glass, and the house wall behind it is half glass and half stone. The glass panes let in the light from the greenhouse, and the stone columns stop and store the heat.

The columns vary quite a bit in size, which in this house was right. They are all 2 feet thick and 8 feet tall. One of the columns backs a wood stove. The stone was laid in the old manner, except with concrete bedding mortar. I used this so I could stack as high as need be on each pillar without starting a new one. I then pointed using white mortar with lime and three parts sand.

I let the stones run pretty wild at the request of the family. If you're going for an organic look, you might as well get some in the stonemasonry, because it doesn't mind being crooked at all, whereas you would have to struggle to make blocks or bricks look funky or organic. With 2 feet of thickness on an 8-foot-high pillar, there is a lot of room to be crooked without losing your structural axis. I laid the stones without any layout. Because the sides of the columns were to have studs of barn timber to frame the glass panes to, the wood was put in first, and I fit the stones to it. This can be seen in FIG. C-1 in the Color Section.

The Kellet house has an interesting backup wood furnace. It is outside, in a separate structure, buried in tons of sand and insulated. The commercial name for this type of heater is Hasha. It has some notable advantages over indoor combustion. There is no danger of burning down the house or depleting the inside oxygen. There is none of the mess and bringing wood in and ashes out. The outdoor furnace can burn long, ugly chunks of wood that would never work in a stove or fireplace. The fire is burned full tilt and clean, and the heat is stored in the sand. Heat is taken off the sand and into the house through a system of water pipes and radiators.

With a woodburner like this, it doesn't matter a whole lot when you make the fire. You can make a fire today and use the heat next week, as long as the mass of sand is big enough to hold it, and insulated enough to keep it. Anything you decide to burn in the furnace will ultimately add to the system's heat store.

With the hot-water system worked in, this is a very sensible approach to heating. You could add the heat of burning leaves to the sand if you wanted. You can gain the heat from garbage burning. It is very flexible because there is so little chance of the fire getting out of control being buried in sand.

In conclusion, the Kellet's houses are inspirations in family cooperation, in doing a lot with a little, and in making a very livable and cozy space with a nice amount of alternate technology. Incidentally, for those who would like to do something brave and fun in housing, I feel it is worth mentioning that the Kellet's were able to sell their very unconventional first home rather easily with a nice profit. Their lack of modern construction know-how was clearly an advantage.

KENNEDY HOUSE

My present house was built as an experiment to get as much free heat as possible with fairly limited resources. The Kennedy house is a passive-solar, masonry house that incorporates stonemasonry for interior block wall facing and for outside retaining walls, which hold the dirt against the east and west ends of the building. Because I built this structure for my family, I was more willing to try some new things than if I had been building for someone else.

The first odd thing about the house is that it was built completely in winter. Stonemasonry, and construction work in general slow down a lot during the frozen months, so this was when I had time away from the pursuit of money-making work. Building in winter called for a few alterations in the procedures.

Building in Winter

The excavation for this house presented no problems in mid-December. In fact, because the site was in the corner of a big corn field, it was good that there was frozen ground for the delivery trucks to drive across, instead of the usual deep mud. The backhoe was able to break through the frozen crust of ground and clear an area about 30 × 60 feet. It was dug down to a depth of about 4 feet. The footings for the house were then dug down another 2 feet, so the bottoms of the footing ditches were 6 feet below grade. The temperature at that depth, even in December, was well above freezing, so there was no danger of the footing concrete freezing before it could set. The only vulnerable place on the footing was the top, so I spread straw over it while the concrete set a few days.

It is worth noting that concrete supplies some of its own heat. The reaction it goes through produces some heat, and the mass of unfrozen material has some resistance to freezing. So if you pour concrete in a big chunk, it might do all right in a temperature below freezing. A thin crust of concrete will lose its heat much more rapidly, however.

Dry-Laid Block

I had recently finished a 1,000-hour stint of stonemasonry on the Pickering house, so I certainly had no delusions about the speed of stonework. To get a building up quickly (3 months was all I could spare), I knew I would have to concede to block work. I reasoned that I would face the interior block walls with stone after the building was up—on weekends and days off, after we had moved in.

After I made the decision to use block, things went fast. I had 1,200 blocks delivered to the job site for under $600.00 in 1982. The weight was spread out around the building. What a head start compared to stonemasonry. It is not unusual to lift a stone 10 to 15 times in rubble work, before it is finally in the wall. In block masonry, every time you lift one, it counts—and if you have a truck drop them off with a boom, you might only lift once.

I had never laid block before I built my house. I had to laugh at how easy it was on the body. There weren't any 100-pound blocks. None were razor sharp or round-edged and slippery. They were made of something lighter than stone. They were half hollow on top of that. The main thing that makes them so easy, though, is that they are made to stack on each other. They stack so well that you only need 8 inches to balance a wall. With stone (rough), you need 24 inches. The stone is more than twice as heavy, and you need three times as much. Also, because of the rough nature of stones, you need to handle them more often. Fitting rubble takes extra time. Pointing stone is harder than pointing block. I feel safe in saying that a rubble stone wall takes ten times as much work as a block wall that encloses the same space—maybe twenty.

I was able to lay the blocks for the house in five days, working alone. It was fast because I laid the blocks without mortar, shimming them occasionally with scraps of sheet metal to keep the faces on the plane of the string.

Dry-laying block is a bit unorthodox, although it has been gaining popularity in recent years as a result of a special bonding mortar that is smeared on the surface of the walls. The claim is that a surface-bonded wall has five times the tensile strength of a conventionally laid wall. This is easy to believe, because a conventionally laid wall has almost no tensile strength, other than its own weight.

The problem with surface-bonding a wall is the expense of the gumbo. I tried a bag. It was $15.00 for 50 pounds, and the sand was already in it. It is mostly portland cement, sand, and fibers of glass or plastic. Each bag covers about 50 blocks. I had 1,200 blocks to do on both sides, so I needed 48 bags of the stuff, at over $700.00.

I realized that for the same money and a lot less work, I could fill the cores of the block with concrete and rebar. This has to be stronger in tensile than the block bonded on the surface. The cores of the block represent a lot more surface area than the outer face, so I suspect that there is a lot more grab inside the block, especially with one piece of rebar down each set of cores. It was less work because I could have a concrete truck come and chute the mix into the cores of the walls in a mere hour or so. I wouldn't have to carry the stuff or mix it or anything—just guide it down the holes. Mixing and handling 48 bags of SureWall or BondWall is not light work.

The concrete in the cores of the block wall also gave it more mass for heat storage. Because the outside temperature was 5°F when I was laying the walls, I knew I couldn't use mortar if I wanted to. I was able to build in weather that I normally would have been stumped by. Manual dexterity wasn't required, so I laid the block with mittens, and the work kept me pleasantly warm.

As far as I know, no test figures exist for the strength of this type of wall, but I feel certain that in tensile and compressive strength, a hollow-core, dry-laid block wall with cores filled with concrete and re-bar far exceeds a conventionally laid or surface-bonded wall. Naturally, a lot of strength is gained just by making the wall more than twice as heavy with the cores filled. Even filling a hollow-block wall with gravel would make it a lot stronger.

Another reason the block work went fast on this building was because the walls were planned for the particular dimensions of the block. The walls, rather than being a particular dimension, like 50 feet × 20 feet, are 14 blocks × 39 blocks. This means there was no block cutting, and the cores aligned perfectly. Each block came halfway over the tops of the ones below, which is important if you want the concrete fill to slide down the wall cores completely. An 8- × -16 cement block is actually 7 5/8 × 15 5/8 inches. The

extra 3/8 inch is normally made up with a mortar joint. This makes for some pretty awful figuring when you try to lay out a dry wall. A pocket calculator helps a lot if you turn fractions to decimals.

Building the dry block walls was like playing with toy blocks. The north, east, and west walls of the house have no openings, so the block stacking was very straightforward. I added a buttress, or *pilaster,* every 12 feet on the inside of the wall for extra strength. To tie the pilasters into the wall, I needed to break several blocks in half, but this is no problem because corner blocks are made to break in half with a few taps of the hammer.

FIGURE 14-14 shows the beginning of this dry-laid block work. A pilaster can be seen at the center of the far end wall. The strings and batter boards are also visible.

I parged the exterior surface of the walls with a regular mortar mix, as a token waterproofing and adhesive strength gain. Actually, I don't think the parging is necessary, but it doesn't hurt and isn't hard to do.

The ditch for the footing was 2 feet wide, which gave enough space for the 8-inch wall, and a drainpipe and insulation on the outside, all resting on the top of the concrete footing. I was lucky that the ditches didn't collapse before I was able to get the insulation and drain in place.

The 4-inch drainpipe encircles the house at the top of the footing level. The inside slab is completely above that level, so water is drained off before it can appear at floor level. The drainpipe that encircles the wall must have a downhill exit. Around the footing, the drain doesn't need to be on a slope. The pipe itself is high enough (4 inches) to ensure that water will flow out even though it might sit a few inches deep in some low spots along the line. This won't hurt, as long as there is still space in the pipe for water to move over the puddles.

I also used an interior drain, which came through the slab at the lowest point on the floor. This drain is used to drain away any water that might appear at the surface of the slab.

After the block walls were up to the finish height, the cores of the blocks needed to be filled. The driver had to drive the truck up on a pile of dirt to get the altitude needed for the concrete to flow down the block cores. If you try to do this with a building on ground level, it won't work. The truck won't be high enough. The only reason the pour down the cores worked for me is because the wall starts 4 feet below grade. This work was done during one of the frequent and predictable December gray spells. These gloomy periods of weather are predictably above freezing, so I thought the concrete would have a chance to set before a cold snap started.

In case it turned bad out too soon, I had the foam insulation ready to prop against the walls. As it happened, I didn't need to do that, but I mention the possibility, because jobs of this nature do have lots of foam sheets and 2 × 4s on hand. Subfreezing weather doesn't necessarily have to stop construction plans.

To grasp the significance of being able to have concrete poured in the walls, think of it as half of the wall being built in an hour. The concrete delivered cost about $20.00 per ton. With the blocks in place, no concrete forms are needed. The block work that seemed so fast compared to laying stone seemed downright poky compared to pouring concrete.

FIGURE 14-15 shows the second floor being built, The upper floor in this solar house is a conglomerate outbuilding. The second floor contains an artist's studio, a big open playroom, and an office. The actual passive solar house is the lower part. Rather than burying the roof of the one-story building and building separate outbuildings for those other needs, I decided to put the second floor on the solar house. This choice saved having another excavation, footing, and roof for the outbuildings. The two floors are thermally separate. There isn't a stairway, and the ceiling of the lower level is heavily insulated.

Having this extra building above the underground house helps the house in many ways. First, the extra weight on the block walls makes them better able to resist any lateral push from the earth bermed against the walls. Secondly, the upper building allowed the use of cheaper, more available conventional roofing materials. Third, the roof, gutters, downspouts, and drainpipes reroute almost all the water away from the buried walls. The interior of the building is bone dry, not damp like a basement. The roof and drains are gradually creating a very dry condition around the bermed walls.

Typical underground houses have much more of a drainage problem than this one because they don't have a roof that is keeping the house from getting rain down the walls. The bermed walls of the house also tend to keep water moving away from the walls. Another advantage of the piggyback design is that the upper floor further retards heat loss from the lower part. If you're going to gather your outbuildings around your house for extra protection, they will do the most on top.

In the foreground of FIG. 14-15, the foam insulation can be seen standing against the wall, with dirt pushed up against it. Plastic can be seen behind the foam. Actually, there are three 1-inch layers of extruded polystyrene with plastic sandwiched between foam layers for protection from the rough exterior of the block wall and the harsh backfill. Only a single roll of 4-mil polyethylene 20 × 100 feet was needed to give two layers of waterproofing the whole way around the buried part of the building. Unlike the Pickering house, there are no nailers on the outside of these block walls to hold the insulation. I merely propped sticks against it until enough dirt could be pushed against the wall to hold the insulation in place.

Also in the foreground of this photograph can be seen the start of a dry-laid block wall for holding the backfill dirt against the wall. The dry-block wall is 16 inches thick and was built with the idea of adding a thick stone facing. The dry-block wall, with no mortar on the faces or in the cores, was able to resist the force of the bulldozer pushing the dirt against the wall. I'm not recommending this procedure with the block retaining walls, although it has worked for me and I think it is worth mentioning. This is obvious testimony to the major role gravity and friction play in the strength of a wall, and the relatively small role the mortar plays.

In the same illustration, strips of foam can be seen covering the block pillars on the open side of the shelter. I used cut nails to put treated wood nailers on the pillars. Then I used 1 inch of foam between the horizontal nailers, and another sheet to cover the nailers and the first layer of foam. I then nailed 1 inch of poplar siding on over the foam, using nails long enough to penetrate the inch of siding, the inch of foam, and part of the nailer.

I built up the beam across the south opening pillars with fifteen 20-foot 2 × 10s. The individual pieces were light enough for me to lift into place, where I kept overlapping the joints and nailing into another

Fig. 14-14. The beginning of a dry-laid block solar house.

Fig. 14-15. The roof purlins are up and one dry block retaining wall and steps are in. Earth is pushed against the block walls to hold the insulation in place.

Fig. 14-16. The wall is about to be filled with concrete in the cores.

2 × 10 until there was a laminated beam. The second horizontal 2 × 10 that is visible above the beam is a band board, which connects the ends of the floor joists sitting on top of that beam. I wanted to be sure that the beam would not flex on the glass below, so it is rather overbuilt.

The west, dry-laid block retaining wall included cap blocks, which transformed the walls into stairways with 16-inch-deep treads and 8-inch-high steps. At the opposite retaining wall is a steel culvert. I built the retaining wall around the 10-foot-deep pipe to give me a fast and easy animal shelter, with passive solar tendencies. The dirt for the backfill of the house came from a hole near the house, which was created simultaneously. The roof drainage and other drainage in the yard goes to the small pond. The pond is too small for fish or swimming, but it makes a nice froggy place with cattails and dragonflies that can be used to irrigate the garden.

FIGURE 14-15 shows the west end on the upper story to have no openings. Actually, that wall has three openings, but they were cut out of the plywood later so we could work in a more wind-sheltered place.

FIGURE 14-16 shows the house completed. The solar doghouse is evident at the right retaining wall. Above that wall is a 500-gallon water tank that gathers roof water to be gravity-fed to the garden, which is in front of the house. In the center of the glass panels is a photovoltaic generator for electricity from sunlight.

It is evident from the photograph that the floor of the lower level is tucked in and very much sheltered from the elements. Only the front wall of glass is exposed to temperatures below freezing. The floor is well below frost line. The ceiling has 10 inches of insulation and another building above to retard heat loss in that direction. The sides and back wall are well insulated and completely underground. The entire south wall is glass, except for the pillars, and the building is situated where none of the rays are blocked by trees or hills. The depth of the building is only 20 feet, so the winter sun's heat enters deeply into the structure, and has a hard time getting back out. At night, during cold times, insulated shutters swing down from the ceiling and cover the glass to keep heat in. They are made of 2 inches of Styrofoam in a hinged muslin case.

These measures are fairly extreme, but very effective in the pursuit of free heating. As you might expect, the solar performance of this house is very good. It should be, since all the design criteria were aimed at that goal. The house (downstairs) has needed backup heat for only two of five winters. I believe that an automated shutter system would completely eliminate the need for backup heating.

The Kennedy house has a modified brick Trombe wall 4 feet back from the center third of the south-facing glass. This tonnage of dark bricks in black mortar not only holds heat and moderates the temperature swings in the house, it supports the center beams, which support the floor joists. The brick wall, which is built entirely of junk brick in a random pattern, is also a room divider and encloses a large planter. This Trombe wall was the most useful piece of masonry I've ever built because it serves so many functions, is nice to look at, and took only four days and $50.00.

A STONE WALL OF MAXIMUM PRACTICALITY

The brick Trombe wall I built for my house got me thinking on ways to make something like it out of stone, and to make it even more useful. I have developed a plan for a stone structure that you can incorporate into many different floor plans, to give the absolute most for your stonemasonry dollar and hour.

FIGURE 14-17 shows a view, from above, of one of the possibilities for a modified stone Trombe wall. FIGURE 14-18 shows a side view of the same wall. The planter wall is within 4 feet of the tall expanse of south wall glass, leaving plenty of space to move past the structure. If you get it back too far, it won't get hit with enough sunshine. The Trombe wall should be toward the center of the length of the house. This way, morning, noon, and afternoon winter sun will strike the stones and warm the mass.

The wall divides the living space and carries the beams, which carry the floor joists. Floor joists above the wall could sit on the masonry.

The front extension on the wall encloses a large planter. The planter wall is about 2 feet high. An inside planter like this, especially in a passive solar home, has no chance of freezing, and that makes all the difference in the world regarding structural requirements. The 2-foot elevation of light, loamy potting soil adds hardly any pressure on the planter wall.

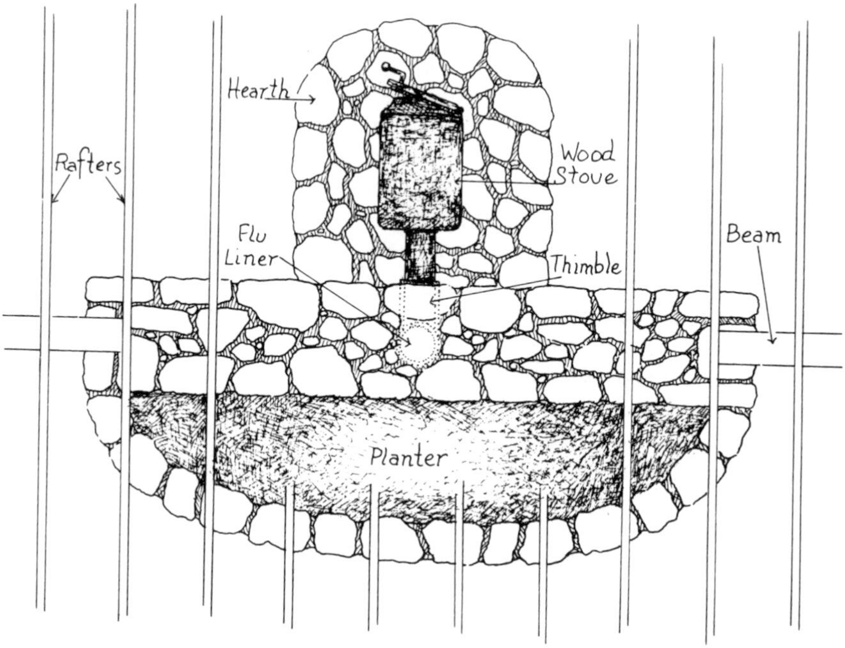

Fig. 14-17. This suggested stone wall is a Trombe, planter, room divider, roof support, chimney, wood stove backup and conversation piece, as seen from above.

Hearth

Rafters

Flu Liner

Wood Stove

Thimble

Beam

Planter

Fig. 14-18. Side view of the most practical of stone structures in a solar house.

Here is an opportunity to build a rubble wall only one stone thick. The inside surface of the planter wall is not seen, so it is a good place to parge with a rich mortar mix. Two parts masonry cement and five parts sand should be good. After you have laid the thin wall, smear the inside surface until it is uniform and tightly sealed. This will do a lot to hold together a thin-walled planter.

Having a planter in a house is a visual treat, and it will keep the air fresher. Theoretically, a sealed house with the right amount of plants and animals would need no outside air at all.

The planter in my house, which is a patch of tropical jungle, is so much easier to maintain than a comparable amount of individually potted plants. The plants are all connected and there are constant changes in the dominance. It is so easy to water against the Trombe wall. The planter contains over 2 cubic yards of a mix of compost, topsoil, clay, and sand. We make the compost from our household vegetable waste, so it gives a good feel to see it back at work, supplying us with a green oasis and oxygen in January. There are also several amphibians living in the planter: two small turtles, a few tiny tree frogs, and a few toads. There is a small pool of water at dirt level, like a lake. In it are a few water plants and snails.

In the suggested Trombe wall/planter design in FIGS. 14-17 AND 14-18, I have gone one step beyond the performance we enjoy with our brick version. The back side of the wall, which would be a cozy, secluded living room, would have a wood stove (preferably a catalytic model with viewing and cooking options) on a big stone hearth. The stove would vent its exhaust through a chimney in the wall.

I think these combinations fit well together because the living room, where the stove or fireplace is, would mainly be a nighttime place. So it wouldn't matter that this area, directly behind the Trombe, was not the sunniest of places in the house. The planter on the other hand, is more of a daytime show. Also, it is convenient to have the heating system hooked into a thermal mass. In this case, the stove would be used when the sun wasn't putting many Btus into the wall tonnage.

In an efficient home, any interior combustion device should have its own air intake. A fire breathes a lot of air, and it causes cold outside air to be pulled in from whatever cracks you have around windows and such. It is best to give the fire its needed cold air right by the stove so you don't have to feel the cold draft of the air. Having the flue of the stove in a thick stone wall should keep the pipe from heating the air or soil too much or too suddenly. Having a fire wall behind a stove, a hearth on a slab under it, and the stove pipe going through a masonry chimney should be about as safe as you can get with wood heat in a house.

What I'm advocating here is a piece of furniture for a modern home. It would have a lot of the feel of a fireplace. An inside stone wall is a good place for fancy, decorative masonry. Trombe walls are good places to show off, because they are not apt to get covered with bowling trophies and stuff that makes folly of the mason's efforts. If someone pays a mason $15.00 per square foot for an artistic wall of stone, it seems silly to cover that surface with pictures, knick-knack shelves, dead animal heads or anything else, yet it happens all the time. A true masonry Trombe wall needs to be free of clutter to function well.

To keep the planter watertight, trowel the inside walls very smoothly with a mix of one part masonry cement and two parts sand, put on very wet. You can tell from the looks of it when you've got the grout tight enough. All the little bubbles and holes will start to be engulfed in the cement paste, and you can get a surface as smooth as plastic. Fine sand is good for this purpose. You can sift some regular sand to get fine sand. I have parged water tanks like this that have held water for four years. I have tried to seal other ones with products like Thompson's Water Seal, with no luck at all.

Concrete has the potential to be as waterproof as anything you can add on top, like paint. If you have ever poured concrete on plastic, and then pulled it up later, you know how slick the surface is that was formed on the plastic. Water won't go through that tight finish. Our planter only suffers a temperature swing of 20°F at most, so the parging mortar has a good chance of long life. An outside planter here would endure a temperature swing of 120°F, which can put the mortar through some changes.

Our planter is a sealed masonry container, with a couple inches of gravel on the bottom. I'm not sure the gravel was necessary. This is not very sophisticated, but we pay attention to the plants to tell what water is needed. There is no drain in the pot. You could leave a hole at a low point, so if you oversoaked it, it would make your floor wet instead of drowning the plants.

If you are going to go to the trouble of doing good stonemasonry, why not have the stones do you some good besides being nice to look at? The stonemasonry furniture of FIGS. 14-17 AND 14-18 can moderate the interior climate and lessen the temperature swing; it can offer a climate and environment for plant life. It can absorb and store the direct heat from the sun; it can store the heat from a fire. It can contain a safe chimney; it can support the roof; it can create a room divider. It can offer safety from the danger of a wood stove, and be a delightful conversation piece.

This plan was developed in answer to a frustration I often feel with my kind of work, which is that most of what people have me build for them is totally unnecessary and imminently impractical. A person likes to feel that their work has a point other than to pay the bills. This proposed solar wall would fill that need, and in my opinion, this is what people should be having stonemasons build instead of fireplaces. Of course, in an existing house, a solar wall like this might not work. A thing of this nature should be designed into a new house.

IN CONCLUSION

If you haven't built yet and are planning a solar house, consider these options before deciding on wood heat as a backup. You can get your backup heat by increasing the solar heat storage, as much as by adding fire.

Most solar house designs could easily add more solar gain and insulation, but it is seldom the direction of a builder's thinking. If you don't have a movable insulation scheme, for instance, you aren't getting the full performance of the building. An automated window insulation system sounds too far-fetched for builders to tackle, yet they don't hesitate to install furnaces with ductwork, fans, and thermostats, which are much more complex than the movable insulation. People just aren't familiar with it yet, but systems are available that would automatically insulate your glass substantially when the voltage dropped to a certain point on a small photovoltaic panel, and would pull the insulation away again when the voltage came back up with the sunshine.

If your house is of the earth-sheltered passive variety, you will probably be getting 90 percent or more of your heat from the sun (unless you live in Vermont). When it is that close to 100 percent, an automatic window insulation system could make the difference. Here's what I mean by the automatic insulation. Suppose the sun is shining on a cold winter morning, so when you go to work, you open the drapes or shutters to let the sun into your solar house. An hour later, the sky becomes very cloudy and the house begins to lose heat through the glass, instead of gaining it. The family can't come back home to put the insulation back on the glass, but an automatic system would do that. A manual override on an automatic shutter system would be used for viewing out the glass, or for when there was enough heat stored. Complex as it might sound, this is certainly not beyond what most Americans take for granted in household technology. If the automatic movable insulation system could mean the difference between needing an expensive backup heat system or not, then you should weigh the cost of the added solar feature against the cost of the backup. Designing a system of automatic glass insulation would be best done from the onset, rather than as a retrofit on an existing house.

Sealing up leaks in a solar house can further minimize the need for backup heat. The irony of a wood stove in a very tight and well-insulated house is that the stove and chimney become the worst offenders for thermal leakage. A chimney will lose heat when it isn't being used. The metal or masonry will conduct heat right out the roof, even if there is no air moving through the chimney. In a conventional house, this is not a significant factor, but in a superinsulated modern house, it can be a more important factor. I have heard of many solar houses that claim to need only two or three fires each year for backup heat. These places might need no backup if they took out the chimneys.

Reducing the amount of toxic material in a house will decrease the need to introduce fresh cold air from the outside. Formaldehyde is particularly offensive in a supertight house, and it is emitted by a variety of common construction and household items. One sheet of new plywood makes the inside of a tight house stink. The effects of direct sunshine on plastics can complicate the toxicity of interior air. Natural materials go hand in hand with solar heating. Sun striking wood, stone, wool, bricks, glass, cats, and people is fine. Synthetic carpets, drapes, and couches, however, will fade from the sunlight, and slowly poison the air while they fade. Cigarette smoking makes an obvious demand on a ventilation system and should be done outside, if at all.

For a supertight, superinsulated, or solar house that needs only a modicum of extra heat, electricity is best. In fact, an electric range should be able to handle the demand. There is no depletion of interior oxygen from electric heat, because the combustion has taken place at the power plant. The electric heat also produces no toxic fumes to eliminate. If a passive solar house starts to feel chilly during a gray spell in February, go on a big bread-baking binge.

That sort of awareness—baking when the house needs heat rather than when it is too hot—will go a long way to make a solar, or any other kind of house, work better. In a solar house, you leave the dirty bath water in the tub until is has given up its heat to the room. Then you drain it out of the house, unless the planter is thirsty. Solar heating is basically about more awareness, not more technology.

BIBLIOGRAPHY

BOOKS

Allen, Edward. *Stone Shelters.* Cambridge, Mass.: M.I.T. Press, 1969.

Allen, George. *As Seen from the Mason's Scaffold.* La Porte, Ind.: Allen Press, 1943.

Baker, Prof. Ira. *A Treatise on Masonry Construction.* 10th ed., New York: Wiley and Sons, Inc., 1914.

Boericke, Art and Barry Shapiro. *A Guide to the Wood Butchers Art.* New York: Simon & Schuster.

_____. *Handmade Houses.* New York: Simon & Schuster.

_____. *The Craftsman Builder.* New York: Simon & Schuster.

Bruyere, C. and R. Inwood. *Country Comforts.* New York: Sterling Publishing Co., 1979.

Daniels, F. *Direct Use of the Sun's Energy.* New Haven, Conn.: Yale University Press, 1964.

Dechere, Rene. *Les Huttes du Perigord.* 1981.

Dezettel, Louis. *Masons and Builders Library.* Indianapolis: Howard W. Sams & Co., 1972.

Feininger, Andreas. *Stone and Man.* New York: Dover Publications, 1961.

Jones, Sydney R. *English Village Homes.* London: B.T. Batsford.

Kern, Ken, Steve Magers, and Lou Penfield. *Stone Masonry.* North Fork, Calif.: Owner-Builder Publications, 1976.

Kern, Ken and Steve Magers. *Fire Places.* New York: Charles Scribners and Sons, 1978.

Knoop, Douglas and G.P. Jones. *The Scottish Mason and the Mason Word.* Manchester, Eng.: Manchester University Press, 1939.

Leckie, J., et.al. *Other Homes and Garbage.* San Francisco: Sierra Club Books, 1975.

Maldon, Leo D. *How to Build with Stone, Brick, Concrete and Tile.* Blue Ridge Summit, Pa.: TAB BOOKS Inc., 1977.

Mazria, Ed. *The Passive Solar Energy Book.* Emmaus, Pa.: Rodale Press, 1979.

McKee, Harley J. *Introduction to Early American Masonry.* National Trust for Historic Preservation in the U.S., 1973.

Ministry of Labour. *Great Britain Youth Employment Choice of Careers.* London: Charles Birchall & Sons, Ltd., 1949.

Muthesius, Stefan. *The English Terraced House.* New Haven, Conn.: Yale University Press, 1982.

Peterson, Charles, ed. *Building America.* Radnor, Pa.: Chilton Book Co.

Roberts, N.H. *The Complete Handbook of Stone Masonry.* Blue Ridge Summit, Pa.: TAB BOOKS, Inc., 1981.

Shelter. Bolinas, Calif.: *Shelter Publications,* 1973.

Stone, Ralph. *Building Stones of Pennsylvania.* Harrisburg, Pa.: 1932.

Time-Life Books. *Masonry.* New York: 1976.

University of Minnesota. *Earth Sheltered Homes.* New York: Van Nostrand Reinhold Co., 1981.

U.S. Dept. of the Interior. The Secretary of the Interior's *Standards for Rehabilitation and Guidelines for Rehabilitating Historic Buildings.* Washington, D.C.: Government Printing Office, rev. 1983.

Van Dine, Alan. *Unconventional Builders.* Chicago: J.G. Ferguson Publishing Co., 1977.

Vivian, John. *Building Stone Walls.* Charlotte, Vt.: Garden Way Publishing, 1976.

Wade, Alex, and Neil Ewenstein. *30 Energy Efficient Houses.* Emmaus, Pa.: Rodale Press, 1979.

Wampler, Jan. *All Their Own.* New York: Oxford University Press, 1977.

Williams, Clement. *The Design of Masonry Structures*

and Foundations. New York: McGraw Hill Book Co., 1922.

Wright, David. *Natural Solar Architecture*. New York: Van Nostrand Reinhold Co., 1978.

Woodforde, John. *Bricks.* : Routledge and Kegan Paul, Ltd., 1976.

MAGAZINE ARTICLES

Engineer's Journal, Gettysburg Battlefield. U.S. Department of War. Hand written journals of engineers working on stonemasonry at the National Military Park, Gettysburg, by special permission of the Battlefield Librarian.

Mack, R.C., de Teel Patterson Tiller, and J. Askins. "Repointing Mortar Joints in Historic Brick Buildings." preservation brief. Superintendant of Documents, U.S. Government Printing Office.

Kennedy, Stephen. "The Structural Stone Wall." *Fine Homebuilding* Magazine 27 (1985).

"The Mortar." *Technology & Conservation.* Vol. 7, No. 2.

"Building Conservation in Great Britain." *Technology & Conservation.* Vol. 6, No. 1.

"Decay of Stone Monuments and Building." *Technology & Conservation.* Vol. 7, No. 1.

"The Russack System for Brick and Mortar Description." *Technology & Conservation.* Vol. 5, No. 2.

INDEX

INDEX

A

absorptiveness of stone, 4
aesthetic masonry, 211
    color and shade in, 214
    movement in, 214
aggregate, 87
air-entraining, 90
arches, 32
argillaceous stone, 2
ashlar masonry, 32, 204

B

banker mason, 208
barbecues, 218
basalt, 3
base (fireplace), 151
batter boards, 37, 67, 70
beauty of stone, 6
bedding mortar, 92, 194
binder, 87
bondstones, 61
bracing stakes, 42
breastworks, 26, 27
brick laying, 216
British thermal units (Btus), 223
building codes, 5
    chimneys, 143, 144
    fireplace, 155
bumpy stones, 59

C

cairns, 215
calcareous stone, 2
cap stones, 18, 19, 63, 72
catalytic combustors, 139

cement, 85
    masonry, 91
    natural, 87
cement butter, 174
chalk box, 33
chimneys
    building codes for, 143, 144
    clean out openings in, 149
    fireplace, 157
    footing for, 147
    foundations for, 143
    height of, 146, 148
    stone, 141
    stone facing, 132-136
    troubleshooting for, 157
    wall and flue liner for, 143,
        147
chimneys and fireplaces,
    137-159
cisterns, 68, 208
color and shade, 214
combustion chambers (see fire-
    places), 155
completion times, prediction of,
    11
concrete, pouring of, 178
concrete block walls
    dry-laid technique for, 239
    stone facing for, 120
cones, 211
corbelling, 156, 211
cornerstones, 63
costs, 9
courses, 15-19
    connection of, 56
culverts, 79

cut stone, 32, 205
    drafting margins for, 207
    technique for, 60

D

dams, 73-84
    considerations, 73
    constructional details for, 79
    culverts in, 79
    environmental considerations
        for, 74
    failure of, 82
    height of, 75
    level and straight, 77
    lining inner walls of, 78
    one-hour, 75
    permanent vs. temporary, 74
    power potential of, 83
    pressure against, 78, 80
    proper heights for, 73
    slowing water flow into, 76
    two-day, 77
    waterfalls over, 81
diabase, 3
diagonals, 34-36
domes, 208
drainage, 183
dressing, 135
    tools for stone, 206
driveway piers, 218
droves, 208
dry-laid technique, 45
    mortared stonework, 127
    steps in, 166
    walls of, 13, 30, 45
durability of stone, 4

E

ecological awareness, 8
end pieces, 17, 54
English-to-metric conversion, 34
environmental considerations, 74
excavation, 38

F

facings, 120-136
  added strength and mass from, 123
  chimney, 132-136
  dry-laid look in mortared, 127
  exterior wall, 126
  footing for, 120
  height of, 130
  inside wall, 122
  layout for interior, 128
  material required for, 124
  openings in, 124
  protective, 122, 131
  stone sizes and shapes in, 130
feathers, (see quarrying)
fill, 55
fire resistance, 4, 131
fire-stops, 157
firebox assembly, 154
fireplaces, 138, 149
  building codes for, 155
  chimneys for, 157
  corbelling in, 156
  dimensions of, 153
  expansion space in, 156
  firebox assembly in, 154
  hearth construction, 155
  lintels and *mantles over, 151
  masonry, 140, 150
  parts of, 151, 152
  Rumford design, 151
  Russian, 139
  steel-lined, 150
  thimbles in, 157
  troubleshooting for, 157
  wood stove conversion of, 142
flagstones, 32, 173-187
  coatings for, 186

drainage in floor of, 183
finding, 173
floors of, 174
laying out for, 176
leveling bed for, 174
mortared floor of, 177
mortarless, 174
pointing and finishing of, 185
pouring concrete bed for, 178
size of, 173
vertically laid, 186
flue liners, 143
  repairs to, 158
footings, 30
  batter boards for, 37
  determining amount of stone needed, 41
  drywall, 45
  excavation for, 38
  layout for, 33
  rubble-stone, 39
  slope built, 39
  tools for, 33
  use of gravel in, 41
foundations, 15, 97-119
  chimney, 143
  construction details for, 107-119
  layout for, 109
  openings in wall of, 111
  repair of collapsed, 193-199
  stone pier, 97-107
  undermining footings in, 108
  underpinning of, 112

G

gabbro, 3
granite, 3
gravel, 41
  see also, mortar
grout, 173

H

head, 83
hearths, 155
houses, stone, 220
  dry-laid block in, 239
  examples of, 226-239

slip-form construction, 221
solar heating of, 226
underground, 228
winter-built, 239
hydrated lime, 85
hydraulic cement, 85

I

igneous stone, 1
imperfectly shaped stones, 49
in-sloping stones, 51-54
internal bonding, 55

J

joints, 52, 208

L

layout, 33-44
  bracking stakes for, 42
  straight stone wall, 43
  tools, 33
leveling, 18, 36, 38
lime and cement, 85
lime-sand mortar, 86
lintels, 151
  charring on, 158

M

mantels, 151
masonry cement, 91
mass, 7, 223
metamorphic stone, 1
metric-to-English conversion, 34
miniature stone wall, 20
mortar, 15, 85-96
  air-entraining of, 90
  bedding, 92, 194
  chart for mixing, 196
  factors affecting setup of, 194
  mixing, 92, 93
  pointing, 94, 95
  repointing, 158
  types of, 91
mortared walls, 30
mortarless rubble stone construction, 11, 45-72
  basics of, 45
  bondstones in, 61

building ends in, 54
cistern building with, 68
corners in, 50
errors in, 58
filling against, 55
fitting stones in, 50
footings for, 45
forces at work in, 60
ideal assemblages for, 47
imperfectly shaped stones, 49
plumbing and leveling, 49
precepts for, 54
sizes of stone for, 48
sloping stones in, 51
stone shapes for, 47
terminology of, 63
tie stones, 62

N

narrow stones, 57
natural cement, 87
noncatalytic high-tech stoves, 139
nonsquare buildings, 208

O

1B stone gravel, 89
one-hour dam, 75
openings in stone walls, 111
out-sloping stones, 51-54

P

passive vs. active gain, 224
pavers, 174
pellet-burning stoves, 140
planters, 218
plug, (see quarrying)
plumb bob, 33
plumb lines, 18
pointing, 10, 94, 95, 197
    flagstones, 185
    mortar mix for, 197
portland cement, 86
pozzolan, 86
predicting completion times, 11
proximity of stone to site, 3

Q

quarrying, 3, 205
    tools for, 204
quicklime, 85

R

ramps, 28, 172
repointing, 188-203
    curing time for, 200
    dressing down mortar for, 200
    tips for, 201
resale value, 8
retaining walls, 27, 30
    backfill against, 70
    capstones on, 72
    failure in mortarless, 67
    failure of, 188
    fine points in, 67
    footing for, 44
    gravel in, 65
    layout for, 44
    mortarless rubble, 63-72
    repair of, 188-193
    sloped, 71
    steps over, 168
    stone facing concrete block, 126
    thickness of, 70
    weep holes in, 69
retempering, 90
risers, 56
roofing, working under, 11
roofs of stone, 209
round structures, 32
rubble-stone footings, 39
    filling trenches for, 40
ruins, 29
Rumford fireplace, 151
Russian fireplace, 139

S

sand, (see mortar)
scaffolding, 159
sedimentary stone, 1
shaping stones, 60
shimming stones, 59
siliceous stone, 2
slaked lime, 85

slip-form stone house, 221
slope built footing, 39
sloped walls, 71
smoke chamber, (see fireplaces)
smoke shelf, (see fireplaces)
solar heating, 223-247
    energy efficiency, in, 224
    passive vs. active gain in, 224
    stone houses using, 226
    use of Trombe walls in, 225
squaring, 34, 35, 36, 38
Spanish bovedas, 211
staking, 40, 42
standing stones, 215
steel-lined fireplaces, 150
steps, 32, 160-172
    constructional details for, 170
    determining number of, 162
    dry-laid, 166
    guide strings for, 166
    measuring stones for, 169
    mortared, 169
    moving stones for, 169
    retaining walls featuring, 168
    rise and run in, 161
    rise and run in uneven, 165
    tread depth of, 162
    uphill, 160
    using small stones in, 172
stone
    building codes for, 5
    bumpy, 59
    buying, 23
    chemical classification of, 2
    chemical variations and impurities, 2
    dressing, 206
    finding usable, 23
    foraging trip for, 26
    gathering and stacking of, 15-32, 15
    good construction qualities of, 3
    narrow, 57
    out- and in-sloping, 51-54,
    practical applications of, 12
    preliminary information for, 1-14

reasons to build with, 6
rough shaping in, 60
shapes of, 30, 31, 32, 47
shimming, 59
transporting of, 22
triangular, 56
types of, 1
usable types of, 24
using ramps to move, 28
utilizing available sizes of, 48
wedge, 57
weighing, 22
weight of, 24, 25
when to gather, 29
working with available, 4
working with existing, 215
stone piers, 97-107
   checking height and level of,
   106
   determining corners for, 102
   footings for, 97
   laying, 104
   laying out footings for, 101
   tapering of, 105
   use of mortar in, 105
stone walls, 27
   elements of, 14

forces at work in, 60
layout for straight, 43
miniature practice , 20-22
mortarless, 45
repair of collapsed, 193-199
surfaces of, 46
stone-facing, 32
stonemasonry
   cement for, 91
   philosophy of, 5
   special aspects of, 204-222
   temperament required for, 8
   weather conditions and, 9

T

therapeutic value of
   stonemasonry, 8
thermal mass, 223
thimbles, (see fireplaces)
throat, (see fireplaces)
tie stones, 15, 62
transporting stones, 22
trap, 3
triangular stone, 56
Trombe wall, 225
   construction of, 244-246
trullo houses, 208

two-day dam, 77

U

underpinning, 97, 112

V

vaults, 208, 211

W

waterfalls, 81
weather, 9, 10
wedges, 57
weep holes, 69
weight, 22
wicking, 104
wood burning, 137
   catalytic combustors in, 139
   dangers of, 138
   improving efficiency of, 140
   noncatalytic high-tech stoves
   in, 139
   pellet-burning stoves for, 140
   processes of, 138
   supplemental heat from, 140
   technological advances for,
   139

Edited by Suzanne L. Cheatle

Other Bestsellers From TAB

☐ **FRAMES AND FRAMING: THE ULTIMATE ILLUSTRATED HOW-TO-DO-IT GUIDE—Gerald F. Laird and Louise Meière Dunn, CPF**

This illustrated step-by-step guide gives complete instructions and helpful illustrations on how to cut mats, choose materials, and achieve attractively framed art. Filled with photographs and eight pages of full color, this book shows why a frame's purpose is to enhance, support, and protect the artwork, and never call attention to itself. You can learn how to make a beautiful frame that complements artwork. In this clear concise reference, the authors describe for you the procedures involved in picture framing. 208 pp., 264 illus. 8 pages full color.

Paper $12.95   Hard $19.95
Book No. 2909

☐ **HOME PLUMBING MADE EASY: AN ILLUSTRATED MANUAL—James L. Kittle**

Here, in one heavily illustrated, easy-to-follow volume, is all the how-to-do-it information needed to perform almost any home plumbing job, including both water and waste disposal systems. And what makes this guide superior to so many other plumbing books is the fact that there's plenty of hands-on instruction, meaningful advice, practical safety tips, and emphasis on getting the job done as easily and professionally as possible! 272 pp., 250 illus.

Paper $14.95   Hard $24.95
Book No. 2797

☐ **THE COMPLETE BOOK OF BATHROOMS—Judy and Dan Ramsey and Charles Self**

Simple redecorating tricks . . . remodeling advice . . . plumbing techniques . . . it's all here. Find literally hundreds of photographs, drawings, and floorplans to help you decide exactly what kind of remodeling project you'd like to undertake; plus, step-by-step directions for accomplishing your remodeling goals. It's all designed to save you time and money on your bathroom renovations! 368 pp., 474 illus.

Paper $15.95   Book No. 2708

☐ **EMERGENCY LIGHTING AND POWER PROJECTS —Rudolf F. Graf and Calvin R. Graf**

Literally packed with practical advice and how-to-be-prepared solutions, this is a handbook every homeowner should have! You'll find a wealth of information on energy-efficient lighting, security systems for home and business, tips on commercially available products for home and camp power generation, plus a wide range of easy-to-construct and highly useful power projects! 224 pp., 188 illus.

Hard $18.95   Book No. 1788

☐ **HOW TO PLAN, CONTRACT AND BUILD YOUR OWN HOME—Richard M. Scutella and Dave Heberle, Illustrations by Jay Marcinowski**

After consulting the expert information, instruction, and advice in this guide, you'll have the basic understanding of house construction that you need to get involved in all the planning and construction particulars and pre-construction choices entailed in building your home. Best of all, by learning how to make these decisions yourself, you can make choices to *your* advantage . . . not the builder's. 440 pp., 299 illus.

Paper $14.95   Hard $19.95
Book No. 2806

☐ **WORKING WITH FIBERGLASS: Techniques and Projects—Jack Wiley**

With the expert instruction provided by this guide, you can use fiberglass to make model boats, flower pots, even garden furniture and hot tubs at a fraction of the commercially made price! These complete step-by-step instructions on laminating and molding make it simple to construct a wide variety of projects—including projects not available in manufactured versions. 224 pp., 252 illus.

Paper $11.95   Hard $19.95
Book No. 2739

☐ **HARDWOOD FLOORS—INSTALLING, MAINTAINING, AND REPAIRING—Dan Ramsey**

This comprehensive guide includes all the guidance you need to install, restore, maintain, or repair all types of hardwood flooring at costs far below those charged by professional builders and maintenance services. From details on how to select the type of wood floors best suited to your home, to time- and money-saving ways to keep your floors in tip-top condition. 160 pp., 230 illus. 4 pages in full color

Paper $10.95   Hard $18.95
Book No. 1928

☐ **THE WOODTURNER'S HANDBOOK —Frank W. Coggins**

Here's your source for step-by-step techniques for making a whole range of turned wood objects, gifts, accessories, toys, tools, utensils, and more! More than 220 exceptionally helpful photographs and drawings provide show-how guidance . . . expert tips on choosing, using, and maintaining a wood turning lathe . . . and practical advice on the types of woods for turning, and how to make your own cutting tools. 224 pp., 233 illus.

Paper $12.95   Hard $21.95
Book No. 1769

Other Bestsellers From TAB